The Best Kept Secrets for Winning SCHOLARSHIPS

By

Brenda S. Smith

BEST KEPT SECRETS FOR WINNING SCHOLARSHIPS
by Brenda Smith

Printed in the United States of America

www.bestkeptsecretsforwinning.com

ISBN 0-9726370-0-1

Book Design by Janet Long

Dedication and Acknowledgement

This book is dedicated to my family. To my husband, Tommy, son, Geoffrey, and daughter, Leslie, thanks for all your love and support. A special word of gratitude to Geoffrey and Leslie, who helped prove my method of scholarship hunting is very successful. To other family members, especially Patricia, Steve, Valerie, Brian, Marta, and friends, thanks for your enthusiasm and encouragement. Every suggestion was appreciated and contributed to my dream of publishing *The Best Kept Secrets for Winning Scholarships*. To my parents, thank you for happy childhood memories, including an extraordinary, 6,000 mile, two-week trip that covered 14 states, Canada, and Mexico at 90mph. The scenery was beautiful, a little blurry, but still beautiful. What a miracle. Despite bears in Yellowstone, flash floods in Arizona, and a blowout on Pike's Peak, we returned safely and lost the dog only once.

Also, I would like to express gratitude to the foundations, businesses, individuals, colleges, military, PTAs, churches, schools, and many other organizations that sponsor scholarships and low-interest loans to help students. Without your generosity, the students could not reach for the stars and grow with an education.

Introduction

Searching for scholarships and financial assistance is a challenge. To be successful, you must discover and follow the strategies of *The Best Kept Secrets for Winning Scholarships*. This book details step-by-step plans for winning scholarships. It is your road map to find and win scholarships to fund your college education. How to search online and offline is important. How to be prepared, stay organized, and watch those deadlines are essential. Preparation and attention to details are very important. Also, some good luck would not hurt. Preparing for a winning scholarship has additional requirements. They include good grades, adequate SAT or ACT scores, extracurricular activities, and job experience, volunteer or paying. By planning ahead and discovering my secret scholarship tips, you will experience the opportunity of a lifetime. You will receive free money for school and graduate from college without being in debt. This resource book is designed to help you fund your expenses for a college education.

Winning scholarship money is a long process that requires a lot of work and dedication. Locating scholarships that you are qualified for is only the first step. Next, you have to win those scholarships by impressing the judges with an outstanding application package, complete with essays, letters of recommendation, and other supporting documents. This book not only helps you learn how to find scholarships, but it guides you through the process of application. Learn how scholarships fit into the financial aid package as a whole. Discover how to search for scholarships both on the Internet and out in the real world. Find out how to create an application package that makes an impression. Your chance of getting a scholarship is best if you apply the right way at the right time. Also, understand how to avoid the scholarship scams that are so prevalent. Follow the tips given in this book, and you will be ahead of the competition. You are a winner with my secrets of finding and winning scholarships. As long as you work hard academically, participate in extracurricular activities, and present a dynamic student package, your scholarship and financial aid search will be first place. This book is designed for students who need scholarships to help pay for higher education. *The Best Kept Secrets for Winning Scholarships* has the answers to make your search successful. Now, get ready, get set, go for the money.

A college degree can provide you, the student, with many opportunities in life. A college education can mean better knowledge, greater potential, more job opportunities, and more money. Is college worth it? Would you be better off financially if you went to work from high school? According to census figures, college is definitely a good financial investment. In fact, an undergraduate degree is probably the most secure high-return investment an American can make. It is important to remember that most individuals with undergraduate and/or graduate degrees earn substantially higher incomes than do individuals without higher education degrees.

So, in addition to earning a degree, college graduates have the opportunity to earn far more money during their lifetime. College graduates on the average, earn $600,000 more in a lifetime than high school graduates. Investing money in a college degree is much safer than the stock market. As a college graduate, you're almost certain to earn much more than you would with only a high school diploma. For most people, the investment is a sure thing, unless you decide to pursue a low-wage career. Besides, in the real world, would your parents give you a hundred grand to invest in the stock market? I doubt it. Mine didn't.

Twenty years ago, four years of public university cost less than a new golf cart. Now, the price tag is more like that of a new car. The cost of attending a public four-year university has shot up 63 percent in the last decade, even after adjusting for inflation. The legislature has decided to pass more of the cost of education on to students and away from the state according to representatives for the various states' higher education boards.

Tuition costs are increasing nationally; college has become less affordable for the poorest Americans at a time when changes in the economy make a college degree more essential than ever. In 1980, a year's tuition at an American public four-year university represented about 13 percent of the annual income of poor families. By 2000, it was 25 percent. The average annual cost of tuition and fees at a four-year school is $3,385 per semester and rising.

It is the dream of every parent and college-bound student: Win that magical scholarship that pays for everything. It could happen but not very likely. College scholarships come in many sizes and in hundreds of different categories, but hardly any will pay the full freight of a four-year college education. Meanwhile, finding the right scholarship requires following the suggestions for the best scholarship search online and offline, as well as making the best presentation possible of your application package.

To get started, first, do your homework. Motivate yourself by reading every word in every chapter. My students and I have had very successful scholarship hunts; some have never paid a penny of tuition. Start early. Begin looking at college materials after the sophomore year. Visit college campuses. Continue involvement in extracurricular activities, church, volunteer opportunities, and work part-time. Don't forget to maintain those grades. Without good grades, you won't need to search for scholarships and financial assistance. Usually the student needs to maintain a "B" grade average or better, score well on college entrance exams such as the SAT or ACT, and excel in some extracurricular activity or music or athletics.

Scholarships come in different packages. Start by selecting a college that prizes academic excellence as much as you do. Not all scholarships come in the form of outright grants. Many scholarship programs include discounted tuition that comes as part of a preferential package from colleges seeking to shape their entering class with top students. The number of "trustee" or "president's scholarships" offers a sign of how strongly the alumni feel about the school. An excellent school, financially speaking, suggests that its graduates think they were well rewarded in their careers, probably by some of the school's alumni. It is nice to get confirmation that your choice of school translates to a head start on your future. Many colleges are proud of the fact that their alumni have a network to help each other after graduation.

Many schools do not offer aid packages to cover all of a student's recognized financial needs. The difference between a student's need and the school's award is called the "gap," which can be as much as 35 percent below the estimated need. For some schools the "gap" is narrowed if the student has achieved high academic or other honors.

Colleges also offer specific scholarships for special achievements and general excellence. Twenty percent of these types of grants come from the assets of the individual colleges and universities. Usually this particular form of grant is based on scholastic excellence. Scholarships for music, art and athletics are clearly defined. Academic awards are based on a combination of a student's SAT score and his or her academic ranking as compared with classmates. Work to achieve that goal.

Scholarships from corporations, community groups and religious organizations abound, if you know where to find them. It's not easy, but I can help you. By following my recommended scholarship search methods, you will attain financial success. Take advantage of the free web search engines. Input your student profile and verify if you are qualified to pursue the scholarships presented. Watch the PTA bulletins and counselors' announcements of available scholarships. Read the paper for articles on scholarships or contests for financial assistance. Visit the nearest Funding Information Center to list the area's foundations that offer scholarships to local students. Spend hours in the public library reviewing the thousands of foundations that offer financial assistance of various kinds. Check the eligibility. The best bet is to analyze your strengths and join groups that promote such talent. Only contact the ones that match your interests and academic qualifications.

When searching for scholarships, part of the challenge is to follow the best advice you can find. Another part is to ignore advice that's misguided. Students with all sorts of talents and interests win scholarships. In fact, many of the best scholarship programs are designed for students who devote their time to such diverse fields as music, the arts, foreign languages, community service, science, leadership, writing, and oratory, to name a few.

Just as the letter grades on your transcript aren't a prerequisite for winning scholarships, varsity letters earned on the playing field aren't a requirement, either. Although a few merit-based contests are targeted at athletes who excel on and off the field, for the vast majority of contests, athletic ability is just another element of a student's record, equivalent to having a skill in any other field.

Scholarship contests demand a different approach than the college admissions process, because contests are characterized more by direct head-to-head competition.

While college admissions officers compare students primarily to a standard, scholarship contests directly compare students to one another. Because of this heightened competitive environment, students who devise creative techniques to make them stand out from the crowd have a distinct advantage. An outstanding scholarship application package will be the advantage.

The reality is you will not rely on any one form of financial aid. You will probably get a combination of loans, grants, private money, and scholarships. Staying on top of all the help that is available is your best approach. While there is no guarantee that all the resources listed in this book will be available when you need them, or on the same terms, they are the most current.

Applying for scholarships is a numbers game. A variety of factors outside of your control can affect the outcome of any given contest. Only by applying for large numbers of scholarships can you minimize the effects of such factors. Every scholarship contest has biases. This is not to say that scholarship judging is unfair. Rather, it's just that each scholarship foundation or organization is looking for students with particular qualities. The ideal application for one particular scholarship contest may place you out of the competition for another. Because of these biases, it's essential to define each contest's ideal applicant and to develop a strategy that emphasizes personal attributes consistent with that definition. What you do after you decide to apply for scholarships is just as important as what you did before. This holds true regardless of how little time you have left until a particular scholarship application is due. Content strategies, for instance, increase your chances of winning by adding depth and breadth to your existing record. Likewise, packaging strategies help make your application stand out by highlighting talents, communicating passions and emphasizing potential. Don't fall into the trap of thinking that your chances are slim. What you do now and the quality of your application package can make the difference in winning.

A college education can help you achieve triumph in today's world. Remember, the only thing more expensive than an education is not having one. The challenge awaits - the key is higher education. Ultimately, your success depends on your personal promise, along with the commitment from family members and teaching professionals to promote and advance education. We recognize that the American family is the rock on which a solid education can and must be built. Years of research clearly show that family and community involvement in children's learning is vital. I want to say thank you for your interest in this resource book and for your commitment to education. I hope that you find *The Best Kept Secrets for Winning Scholarships* will be a valuable resource that will make you the winner of thousands for your college education. I wish you the best in your educational and career endeavors and look forward to hearing your stories of success.

Table of Contents

1 How To Experience an Incredible Search Online – Helpful Websites

Paying for college has become increasingly more difficult. In recent years, college tuition has increased at an average rate of 8% each year, far outpacing the rise of inflation. At the same time, financial aid awards from traditional sources like the federal government and the schools have decreased. Winning scholarships from private sources has become an important way for students to make up the difference.

In order to find money for college, you need to take the initiative to search out the scholarships for which you are well qualified. However, locating scholarships is often a time-consuming process. You have to page through gigantic books to look for the scholarships you might qualify for, only to find when you apply that important data has changed since the book was published. With growing popularity of the Internet, that situation has changed. Now scholarship programs are being publicized on the web, making it easier for you to find up-to-date information and enabling you to submit applications quickly.

You are probably aware that the Internet is becoming more of a useful resource for scholarship searches. There is no printed material any more up to date than the Internet, and it is usually free if you have Internet access. You are also able to apply online in some instances. When doing a scholarship search, it is recommended that you have an email account. The search sites will automatically match available scholarships to your personal profile. Also, the website will email updates to the student as new scholarships become available.

An undergraduate scholarship is a form of aid that helps students pay for their education. Unlike loans, they do not have to be repaid.

Hundreds of thousands of scholarships from several thousand sponsors are awarded each year.

Generally, scholarships are reserved for students with special qualifications, such as academic, athletic or artistic talent. Awards are also available for students who are interested in particular fields of study, which are members of underrepresented groups, who live in certain areas of the country or who demonstrate financial need. The best way to search for scholarships is to use a personalized search that compares your background with a database of awards. Only those awards that fit your profile are identified as matches.

Like any guide, this book cannot list every scholarship program. You should also do detective work on your own to find scholarships for which you are uniquely qualified. Fortunately, with the entire Internet at your disposal, that search should not take you very long.

Your first stop should be the websites of all the colleges where you are applying for admission or the college you already attend. College websites are often gold mines of scholarship information, if you know where to look for it. First, check the admissions page, which often lists merit scholarships. Then, go to the student financial aid page, which may also list merit scholarships. Check the page for your academic department to discover if any scholarships for your field of study are mentioned. The website may also contain information about artistic, athletic, study abroad, ROTC, and other scholarships sponsored by the school.

The second step in winning scholarships is to hunt for the scholarships for which you are most qualified. While thousands of scholarship programs are out there, no one will just hand the money to you. You have to search it out. There is no central place where scholarships are advertised; no book or database can claim to list all available awards. There is not one screen to click like magic and all the scholarships appear on a list. Therefore, you must use every resource to search for scholarships to create the most complete list possible of scholarships and financial aid contacts that you have the greatest chance of winning.

Most scholarships search services require you to fill out a personal profile listing your individual characteristics and accomplishments. The program then combs the database of scholarships to find those that you may be eligible for, according to the information that you provided in the profile. Therefore, you should be as accurate as possible when completing the profile in order for the program to result with the best list of scholarships for you. Read the instructions very carefully.

Typically, your scholarship search results will provide a lot of useful information about each scholarship, including eligibility requirements, the amount awarded, the deadline for application, and an address where you can write to get an application form. To request a scholarship application via the mail, write a brief, polite note indicating your name, your year in school, the name of the school you attend, and the scholarship that you want to apply for. Include a self-addressed, stamped envelope for the reply. (Always use the biggest, prettiest, and most unusual stamps for all your postage. The envelope that gets noticed receives special handling.)

Since no scholarship database is complete, you should take the time to search as many of them as possible. Of course, you will find some duplicates, but performing multiple searches is the only way to be sure that you have found all the scholarships for which you are qualified. Besides, the searches are free and don't take very long to complete, so what have you got to lose?

There are several free scholarship databases available online. With more than $1 billion in scholarships, FastWeb is considered to be the largest, most accurate and most frequently updated scholarship database. If you supply an email address, they will notify you when new awards that match your profile are added to the database. You can even submit an electronic application to some of the scholarships listed on FastWeb, saving you time and money. FastWeb also includes a college search and other student information.

The most important thing to remember is not to get lazy. Do not take anything for granted; scholarships are not easy to win and competition is fierce. Getting a scholarship is an important way of helping finance your college education and reducing your dependency on loans. Not to be neglected is the resume-enhancing effect of getting a scholarship, even if it is not a prestigious, nationally known one. So, put as much time as possible into this endeavor; it will pay off.

The power of the Internet can help you discover the secrets of finding and winning scholarships. Do not expect it to be your only source. Use it as another tool to augment your search. Competition is tough, and the scramble for financial aid is brutal, but using the Internet will greatly improve your odds. Seek out as many scholarships as possible and apply, apply, apply!

Helpful Websites and Contacts

www.bestkeptsecretsforwinning.com
4Scholarships.com
www.aasa.org/discover.htm
http://www.aauw.org
About.com
www.acinet.org
www.act.org
www.aflcio.org
www.aflcio.org/scholar
www.airforce.com
www.americorps.org
www.army.tcu.edu
http://www.att.com/foundation/circomm.html
www.bls.gov
www.campustours.com
http://www.cashe.com
http://www.cbefnet.org, Congressional Black Causus Foundation
www.chase.com/leducationfirst
http://www.chcl.org, Congressional Hispanic Causus Institute
www.chronicleguidance.com
www.citibank.com/student
www.cns.gov
www.college4u.com
www.collegenet.com
www.collegeboard.org, College Board Online
CollegeBoardFinancialAidServices
www.collegeEdge.com
collegeclub.com
www.collegeispossible.org
collegelink.com
www.collegeexpress.com
www.collegequest.com
www.collegeview.com

www.collegesavings.org

connect.com

cool-College Opportunities Online

www.collegetown.com

www.ctown.com, Lets you visit an admission office

www.dallasnews.com, sampling list of scholarships recently announced by colleges and organizations in the Dallas-Fort Worth area

easi.ed.gov, Project EASI

www.ed.gov, U.S. Department of Education

www.ed.gov/inits/hope/

www.ed.gov/prog_info/sfa/fafsa

www.ed.gov/prog_info/SFA/Studentguide, Department of Education's student Guide

www.ed.gov-pubs

www.ed.gov/offices/OPE, Office of Postsecondary Education

www.edu.com

Eduzone

Embark.com

EStrong.com:College

www.estudentloan.comwww.excite.com

ExPan from College Board Online

www.fafsa.ed.gov

fastaid.com

www.fastweb.com, Fastweb Scholarship Search

Federal Pell Grants

http://web.studentservices.com/fastweb, Financial Aid Search

www.fic-ftw.org, Funding Information Center

www.finaid.org, The Financial Aid Information Page

www.finaid.org/calculators

www.fenbd.com

freschinofo.com

www.ftc.gov

Fundsnet Online Services

Gale's Encyclopedia of Associations

GoScholarshipSearch
GRIP Scholarships
www.honoring.com
www.irs.ustreas.gov/prod/forms-pubs/
http://iwc.pair.com/scholarshippage/links.html
www.jayi.com/jayi
www.jvd.net/scholarships.htm
www.kaploan.com
www.kaptest.com
www.loanconsolidation.ed.gov
www.loanstar.com, Loan Star Lenders
L'Occitane and the American Foundation for the Blind
MoneyCentralFamilyandCollege
www.mybytes.com
www.naas.org
www.nasfaa.org
www.nasfa.org.est.efc, Receive estimated financial aid contribution figure
National Student Loan Clearinghouse
www.nelliemae.org
www.nhsf.org, Hispanic Scholarship Foundation
www.novakint.com/colleges
www.octameron.com
www.oicc.capnet.star.tx.us, Texas State Occupational Information Coordination Committee
OSAD Scholarship Search, online study abroad directory
Outagency
Patterson's Collegequest
www.pe.net
www.petersons.com
www.phecfld
www.pin.ed.gov
P.L.A.T.O. Free Online Scholarship Search
www.princetonreview.com, The Princeton Review
www.prudential.com

www.rams.com/srn

www.rspfunding.com

www.salliemae.com/consumer, Sallie Mae

www.scripps.com/foundation

www.scholaraid.com

scholarship.com

scholarships.com

www.scholarships.salliemae.com

www.searchbyvideo.com, Search by video, company has more than 400 college videos available through video library

www.signet.com/collegemoney/toc, The Ambitious Student's Guide to Financial Aid

www.soicc.state.tx.us

www.sourcepath.com/caid/, Resource Pathways' College Financial Aid Resources

Southwest Student Services Corporation

Spring's Scholarship Page

www.srnexpress.com

studentadvantage.com

www.studentawards.com, For Canadian Students Only

studentloan.com

www.students-gov

web.studentservices.com

Supercollege.com

www.targetusa.com

www.tgslc.org, Texas Guaranteed Student Loan Corporation

www.thecb.state.tx.us, Texas Higher Education Coordinating Board, (Admission and Scholarship info)

theoldschool.org

http://udallfoundation.org/scholarships/

www.umass.edu/lrrc/

http://unausa.org/programs/

Uni-tel Student Scholarship Program

www.usagroup.com/scholarship/scholarship,htm, USA Group Scholarship Program www.usnews.com, U.S. News Best Colleges

www.upromise.com, Upromise (the way to save for college)

www.usmall.com/college/search, College Find

www.va.gov/education/

http://window.state.tx.us/scholars/mspmain.html, Minnie Stevens Piper Foundation (Scholarship services and info for Texas Colleges)

www.window.state.tx.us, Texas Tomorrow Fund (Savings plan for future payment of college tuition at today's prices)

www.wiredscholar.com

http://www.women

While all of the information listed in this book is accurate at the time of publication, some details are likely to change from year to year. Always check each scholarship's website for the most up-to-date deadlines and other information before submitting an application. Be sure to pay close attention to deadline specifics, such as received by 10a.m. on a particular day or postmarked by a certain date. Watch for postmark deadlines that fall on Saturday or Sunday. Some areas do not offer postal service on the weekend. If you need a postmark on Sunday, find the nearest 24-hour post office. If the website appears to be out of date, contact the scholarship sponsor directly and request a printed application and information material.

Combine your lists of possible scholarships that match your interests and talents with your impressive student profile, and you will be on your way to winning extra money for college.

2 How to Search Offline and Defeat the Competition

Financial assistance opens the door to higher education for many who could not otherwise afford its cost. Financial assistance includes various forms, including scholarships, grants, or low cost loans. Be thorough in your search for scholarships. Many scholarships, particularly college and local scholarships, are under publicized, but these are also the awards that you are most likely to win. Therefore, you cannot just search one computer database or leaf through one scholarship guide and quit. Take advantage of as many sources as you can to build as complete a list of scholarships as possible. If you take the time to perform a thorough search, you are much more likely to receive scholarship money.

While the Internet is an invaluable research tool and can make your scholarship search significantly easier, you will not find every scholarship publicized online. Therefore, you should expend some effort out in the real world also. The first place to start is in your own community, where you will likely find the most scholarships that you are qualified for, scholarships that are least publicized on the Internet. Be aggressive in looking for scholarship money. Aside from the university you have chosen, look for scholarship announcements in the following places:

- Your high school guidance office or college academic department
- Your church, synagogue, or temple
- Local government offices
- The Chamber of Commerce
- Your employer
- Your parents' employers, employers of other family members

- Clubs, athletic teams, academic societies, and other organizations you belong to
- Clubs, unions, professional associations, and civic groups that your parents belong to
- Civic and public service organizations
- Your local library, where scholarship announcements are often posted
- Your local Funding Information Center
- Your local newspaper, where scholarship announcements are often published
- The Boy and Girl Scouts of America
- Associations, particularly those representing your field of interest (There are more than 5,000 associations in the United States.)
- Corporations
- Local PTA bulletins that list the scholarships awarded or offered to the previous year's graduating class
- Local high school counselors receive flyers and applications from area businesses, foundations, and organizations offering scholarships
- State educational agencies
- Colleges (Admissions Office and Departments)
- Ethnic or racial heritage groups
- Local professional organizations
- American Legion, Jaycees, etc.
- Texas Rehabilitation Agency (includes physical disorder, asthma, diabetes, etc.)
- Professional organizations such as American Medical Association or American Bar Association
- Hobbies or interests such as rodeo, fishing, or bowling
- Veterans' benefits - parents who have served in the military
- Service organizations such as Lions and Rotary
- Insurance companies
- Local professional sports teams
- Special talents such as art or drama

Visit the public, college or university libraries in your area. Also, visit the nearest Funding Information Center and check out their website. All of these libraries include directories of foundations, associations, corporations, and miscellaneous institutions that offer financial assistance of various types. There are scholarships, fellowships, grants, and loans described including their dollar amounts and eligibility and what the foundation's interests include. In the library, you should peruse the foundation directory, foundation grants index, source book profiles, national data book, directory of (your particular state) foundations, directory of (your particular county) foundations, corporate foundation profiles, Internal Revenue Service information returns for all (your particular state) foundations, annual reports for all state foundations, periodicals on philanthropy, fund raising, and non-profit management.

The key to winning scholarship money is to find awards that match your talents, skills, academics, and interests and expend your efforts in applying for them. No matter how good a student you are, if you do not match the eligibility requirements for a particular scholarship, you do not have a hope of winning it. It will help your scholarship search efforts if you make a list of all of your achievements, interests, and other qualifications before you start. Creating this personal profile can help. An easy form to use is the Student Activities form included in this chapter for your convenience. Think of scholarship programs that you might have overlooked. The profile will also be useful as you begin filling out applications and writing essays. Any of the following factors could qualify you for a scholarship, include them all when checking the eligibility requirements for the various scholarships.

Besides your high school interests and activities, such as volleyball, basketball, academic decathlon, student council, or any other, be sure to search for scholarships that could be included from your volunteer, community, and, or church involvement. The following list could be helpful.

- Gender
- Race or ethnic background
- Citizenship
- The state, country, and city where you live
- The name of your college, or the name of your high school and the colleges where you applied for admission
- Your elementary and middle school that you attended
- Any disabilities that you have or your parents have
- Your year in school
- Your grade point average
- Your class rank
- Your PSAT/NMSQT score

- Your SAT I or ACT scores
- The field you plan to major in
- Your strongest academic subjects
- Academic awards you have won
- Special academic programs you have completed
- Sports you actively participate in
- Student government positions you have held
- All clubs and other organizations you belong to
- Hobbies you actively pursue
- Artistic talents or pursuits (visual arts, photography, dance, drama, music, or creative writing)
- Your employers, present and past
- Internships you have completed
- Community service and volunteer activities you participate in
- Religious organizations you are affiliated with
- Military service you have completed
- Where your parents are employed and their occupations
- All professional associations, unions, clubs, sororities, fraternities, civic groups, and other organizations your parents belong to
- The colleges your parents attended
- Whether your parents are veterans or currently active in any branch of the Armed Forces or the National Guard

Some scholarship programs do consider financial need when deciding who will be awarded the money. If you have already applied for federal financial aid, you should have a good idea of how much financial need you have. If you have not filed the FAFSA or received the results, consider your family income when estimating your financial need. Most scholarship sponsors will not consider you needy if your parents make a decent income, particularly when you are competing against families who make $20,000 a year or less. Do not apply for scholarships that factor in financial need unless you are sure that you have need and can demonstrate this to the scholarship committee. If you need help making this decision, the financial aid officer at your college can often provide a good reality check.

These are just a few of the eligibility requirements that you are likely to confront when you start searching for scholarships. Other scholarship programs may specify that you have a particular ethnic background or minority status, that you belong to a particular religion, or that your parents belong to the sponsoring group or be employed in a particular profession. If you do not understand all of the eligibility requirements, contact the scholarship's sponsor by phone or email to clarify.

Because you meet all of a scholarship's eligibility requirements does not mean that you are the best candidate for the scholarship. Some scholarship programs are very particular and competitive. These programs usually require the most effort in terms of writing essays, gathering letters of recommendation, and submitting additional materials with your application for the scholarship. Before you decide to apply, evaluate yourself honestly to determine if you are a competitive candidate for that award. Check the profiles of past winners to see if your qualifications are similar to theirs; often you can find these profiles on the scholarship program's website. If the average grade point average of past winners is 4.0 and yours is 3.5, for instance, you are not very likely to win that scholarship. Or if past winners all had community service experience and you don't, you won't make a very good candidate. Other qualities that you should consider include leadership experience, creativity, ability to rise above disadvantages, and even personal character. Compare yourself to past winners in these areas to determine if the scholarship judges would think that you are a good candidate for the award.

Reading over the application form can also give you a clue as to whether you will make a good candidate for that scholarship. You should have a strong response to every question on the application. If you find questions that do not apply to you, then that scholarship probably is not meant for you. Spend your time and effort on a scholarship that fits your talents, skills, and academics more closely.

In your scholarship search, try to find scholarship programs where the competition is not so great. You have a much better chance of winning these awards. For example, it is much more difficult to win a nationally known or advertised scholarship because there is much more competition. Instead of expending your efforts on the large national scholarship competitions, look for local scholarships. Find scholarships aimed at students of your ethnic group or gender. Remember that the more specific eligibility requirements a scholarship has, the less competition there is likely to be for that award. For instance, a scholarship for African-American microbiology students from Santa Barbara, California, will probably have very few applicants; if you fit all of those requirements, then you should definitely apply.

Your scholarship search efforts should have turned up a long list of scholarships for which you are qualified to apply. To maximize your chances of winning some scholarship money, apply for as many of the scholarships you found as you can. You should apply for at least 10 scholarships every year you are in college. Just remember to spend the bulk of your time and efforts on the applications for the scholarships that you honestly feel you have the best chance of winning, which are not necessarily the scholarships with the biggest payoffs.

Applying for scholarships is a time-consuming process. You should only spend your time applying for a scholarship that you have the best chance of winning. Remember to limit your applications to the scholarships where you match all of the eligibility requirements. If one of the eligibility requirements says prefer high school resident of a particular county, then you may apply. If the description says, must be a resident of that county, do not apply. Many students believe that if they take a scattershot approach and apply for each and every scholarship they find, they are bound to win something. Not so. If you do not meet the scholarship's, often rigid, eligibility requirements, your application goes into the trash, literally.

Another very important strategy for your successful scholarship search is to begin at the end of the alphabet. That is, when searching for scholarships in each listing or directory of foundations, begin at the back of the book and work forward. Most students will begin their search from A to Z, but will quit before identifying the foundations that start with letters from the end of the alphabet. This strategy improves your chances dramatically. Most of the time, when you locate a scholarship that you are interested in, you will have to write to request the application packet. To do this you will write a letter. The letter should be short and to the point. You will want to include such things as:

- Your name
- Your grade or the year you will graduate
- Your GPA if the scholarship is academic or you have a strong GPA
- Test scores if they are a requirement for eligibility
- The reason you are eligible for the scholarship
- If the scholarship requires financial need, include a brief explanation of your circumstances

An example of the sample letter requesting a scholarship application follows. It is very important to attract attention to your request, in a positive way. To do this, use the brightest, largest, most unusual, and prettiest commemorative stamps you can find at the post office. People notice beautiful or recently released stamps. This is another way to help get your request or reply processed quickly.

The search process does take time, but the suggestions mentioned will improve your chances greatly while searching for scholarships. There are over 2.5 million awards worth over $21.6 billion that are listed each year in directories and databases. Why should you look for scholarships? The bottom line is that scholarships equate to free money, and not too many people are in the position to turn down free money. Remember not all scholarships are based on grades or test scores. For example, you can get a scholarship because you were born in Texas or just because your dad works for the railroad. You would be surprised to know just how many awards exist for all types of students. So, get busy and get started compiling and investigating a list of scholarships. Soon your mailbox will be filled with scholarship applications, and awards. You will be on your way to paying for college with free money.

Letter Requesting Scholarship Application

Date

Mr. Geoff Chandler, President
National Architecture Scholarship
2020 Marta Circle
Caroline, TX 75149

Dear Mr. Chandler:

This letter is an introduction of myself, Ryan Whittington, and to tell you of my desire to participate in the National Architecture Scholarship program. Also, I would like any information for other scholarship programs available for Architecture students. I have been accepted at Columbia University, for the 2003 fall term. Would you please send me an application form? A stamped self-addressed envelope is enclosed for your convenience.

I would like to thank you and the scholarship committee for supporting college bound students with an opportunity for financial assistance through your generosity.

Sincerely,

Ryan Whittington
5099 Leslie Lane
Skyler, TX 76123

3 How to Design the Winning Student Resume

To build a great scholarship application, you need great "content": the activities, awards, achievements, experiences and credentials that make up the application. Great content, however, does not just happen. During the summer is the perfect time to get busy increasing activities to improve your application contents. If you already excel at something, why do you need to add to your record in that area? Because by transforming your strongest skills into truly standout talents, you will get noticed by the scholarship judges.

To develop your talents:

- Use jobs as learning experiences. Contact organizations, professors or professionals in your area of interest and inquire about job or internship possibilities. If they do not offer those programs, suggest ways you can contribute to their department and ask if they might create a spot for you. Many professionals enjoy serving as mentors.
- Devise your own project. Summer is a great time to pursue an independent project that extends your skills. One science-minded scholarship winner from Alaska used the summer to pursue his own science project on the Exxon Valdez oil spill. Another artistic scholarship winner painted murals on old buildings to spruce up the community.
- Leverage your other summer activities. If you are traveling over the summer, turn your trip into scholarship gold. I know of one scholarship winner who transformed a summer trip to Europe into a scholarship-enhancing endeavor simply by keeping a journal of the trip and turning it into an independent study paper on cultural differences.

- Improve your weak areas. Judges value well-rounded candidates. So although your application spotlight should shine on your strongest areas, you will want to include other credentials that demonstrate a wide range of experience.

To build up your weak spots, serve your community. Many students have told me that they just do not have enough time to do community service work. Why not devote some extra summer hours to a worthy cause? Pick a cause you are interested in, and join an existing program or create your own. If you are someone who cannot get enough of the summer sun, find a community service endeavor that gets you outdoors.

Take a summer course. School in the summer may sound like a major drag, but taking one course at a local community college can be fun. Pick a subject that you have struggled with in school, and use the summer course as an opportunity to develop your competence.

Explore old subjects in new settings. Sometimes subject areas that seem dull or difficult in the school environment are not so bad once you get out of the classroom. Participate in programs, camps, clubs, activities, workshops and conferences to explore subjects that are new to you.

So this summer, look for activities that expose you to things you are curious about. You will learn a lot and meet new interesting people. These experiences can also provide you with great raw material for those upcoming scholarship essays. If you utilize just a small portion of those lazy, hazy, crazy days of summer to pursuing these worthwhile goals, you'll be way ahead when the back-to-school regimen starts up once again.

In the process of writing this book, I have had the opportunity to inspect a wide variety of scholarship applications submitted by many different students - scholarship applications that brought home the scholarship bucks, and plenty of others that did not. When I put these two groups of applications side by side, the differences were like night and day.

Unsuccessful scholarship applications, more often than not, resembled a laundry list of activities, awards, and accomplishments. Although these credentials were often impressive in and of themselves, reading such materials felt like scanning a random conglomeration of facts.

When reading a winning application, on the other hand, I felt like I was actually getting to know the person who submitted it. I came away from each feeling like I understood the core interests, skills and values of the individual.

Why do winning scholarship applications ultimately leave such strong impressions? Quite simply, these applications create vivid portraits of the applicants. Each one of these applications does not just recite accomplishments; it depicts the person behind all of the grades, extracurricular activities and awards. After all, judges award scholarships to people, not to resumes. If a judge feels like he or she knows you, it creates a powerful, personal impression that helps your application become more outstanding.

So how does one paint these self-portraits? First, treat each component of a scholarship application (such as the essay, extracurricular activity list and recommendation letter) as part of a unified whole, not as a separate entity. Just as combining the melody of a romantic ballad with the lyrics of a rap would create musical chaos, if you do not coordinate the message that each part sends, the overall impression you leave is unclear and fragmented.

Second, we must develop the message itself, the core idea that permeates the entire application and powerfully communicates who you are. I call this cohesive message the application theme.

Think of the theme as the main point you are trying to communicate about yourself, the framework that puts all of your activities, interests and credentials in the proper context. Application themes are frequently created around particular activities that you are passionate about, particular interests that fascinate you or career goals you are striving towards.

In my experience, the typical application should have no more than one or two major themes. Any more than this, and you start to dilute the communicative power of your message. If an application has two major themes, then one should be the primary theme, and the other the secondary theme. The primary theme should be the focal point of the application, the primary message you are trying to communicate. The secondary theme builds upon this message by illustrating another important aspect of your life, interests and personality.

In one scholarship application, for instance, the primary theme was built around the passion for writing and the deep appreciation for all forms of communication as tools for solving problems. The secondary theme focused on how the student had already exhibited, and would continue to exhibit, a high level of leadership. The student placed credentials that emphasized these themes in prominent positions at the top of extracurricular activity and awards lists, addressed these themes in the essays and personal statements and chose recommendation letters that also reinforced these key essays. Still included in the application materials is the other information and credentials that did not relate to the thematic message, but you want to show a well-rounded student.

Personal Data Needed to Complete All Applications

To help complete all scholarship applications, use the student activities worksheet as your guide. Update your worksheet when your academic record changes. Also add any new honors, awards, or activities.

- Name and address of college where student has been accepted
- Name of county that college is located
- Name, address, and phone number of the financial director
- All names and addresses of schools attended by student
- High school name, address, phone, fax and six-digit identification number
- Transcript, Conversion of GPA to 4-point Scale
- Class rank and size
- Selective Service; Be sure to register online. Some colleges ask if the male student has registered. This shows responsibility, good citizenship, leadership, etc.
- SAT and ACT scores and Student Score Report which reflects national percentile.
- Copy of FAFSA report (Most universities, and foundations and loan organizations require the student to file a FAFSA report before aid is awarded. The Federal Application for Student Aid form may be obtained online at www.fafsa.ed.gov or write to:
  Federal Student Aid Information Center
  POB 84, Washington, DC 20044.)
- Copy of Student Award (from the university where student will attend, received later)
- Birth certificate

- Latest grade report
- Email address
- Wallet size photos
- Letters of recommendation from counselor, 2 teachers, employer, pastor or adult church friend, volunteer directors, coaches, or a close family friend. To help inform them of your current academic status and activities, give them a copy of your student resume when you ask for a letter of recommendation. Send a note of thanks to each person for the letter of recommendation.
- College housing and tuition costs

You will need the following information on the student, parents, step-parents, and siblings:

- Driver's license number
- Social security number
- Address
- Phone
- Date of birth
- Place of birth
- Name and address of current school sibling or parent is attending
- Full name of parents and siblings
- Highest level of education of parents and name and address of that school
- Student income tax return and W2 form
- Parents' income tax return and W2 form

STUDENT ACTIVITIES

Name ____

Please list significant extracurricular activities, community service, talents, awards, honors, and employment.

Fresh	Soph	Junior	Senior
From _____ To _____	From _____ To _____	From _____ To _____	From _____ To _____

Example:

Fresh	*Soph*	*Junior*	*Senior*
From 1998 To 1999	*From 1999 To 2000*	*From 2000 To 2001*	*From 2001 To 2002*

Awards / Achievements

Award / Achievement	Description	Fresh	Soph	Jr	Sr
Examples:					
Regional Qualifier (Speech)	*1 of 3 chosen for regional team (UIL)*	*Regional*		*X*	*X*
All State Basketball Team	*1 of 5 in state chosen for first team*	*State*		*X*	*X*

Employment / Internships

Job / Title	Your Specific Role	From / To	Hours per Week
Examples:			
Sacker	*Sacked groceries, helped customers*	*6/93 — 5/96*	*10*
Exchange Program	*Spent two months with host family in Spain*	*6/96 — 7/96*	*n/a*

Extracurricular Activities and Leadership Positions

Organization	Your Position / Year	Elected Position (circle one)	Level	Description of Activities	Hours Per Week / Weeks Per Year Fr	So	Jr	Sr
Examples:								
Speech/Debate Club	*Committee Chair / senior*		*Regional*	*Hosted Speech Tourn.*	*3/32*	*4/3*	*6/32*	
6/32Basketball team	*Captain / junior*		*Local*	*Played Forward*				
		yes / no						
		yes / no						
		yes / no						
		yes / no						
		yes / no						
		yes / no						

Community or Volunteer Service

Service / Volunteer Work	Your Specific Role	From / To	Total Hours
Examples:			
Habitat for Humanity	*Helped build houses*	*5/94 — 5/96*	*8*
Hospital Volunteer	*Helped nurses and visited patients*	*6/96 — 8/96*	*36*

Talents / Special Honors

Talent / Honor	Activity	Fr	So	Jr	Sr
Examples:					
Voice	*Sang in high school choir*				
Public Speaking	*Won Rotary Four-Way Test Speech Contest*				

List of Folders Needed to Stay Organized

Most recent Income tax and W2 forms of student and parents

- College Applications
- College acceptance letters
- Official Transcripts
- SAT, ACT scores and report explaining national percentile of student's score
- Student Resume
- Selective Service, if male, print the page when you register online —shows responsibility, leadership, good citizenship.
- Letters of Recommendation
- Estimated housing and tuition costs from the college catalog
- Birth certificate
- Wallet size photos of student
- Copies of submitted scholarship applications
- Copies of letters requesting scholarship or loan information
- Scholarship applications completed and ready to mail
- Medical form with information for dates of vaccinations and/or shots required
- List of most recent scholarships offered the graduating class for the previous year. Most PTA clubs publish a newsletter at the end of the year with the recipients listed with each of their scholarship awards. Also, some PTA clubs publish a monthly PTA bulletin listing scholarship forms that are available from the counselors' office or scholarship websites and deadlines. These bulletins are very helpful and should be used in your scholarship search.

Student Resume

The student resume is a short biographic resume of your life. The resume contains all essential facts necessary to present a realistic description of you, your personality, your aptitudes, your ambition, and your background.

For most students, winning a scholarship is not only a matter of luck. It is the result of tedious research, careful proofing of forms, requesting information early, and putting in the hours it takes to have good grades. An exceptional student resume will add more integrity to the scholarship application package. Scholarships are not reserved for only A students even though your grades will be judged. A scholastic rating in the upper third or quarter of your class, and a high school average of at least 85%, improves the profile as a student worthy of assistance. If your grades are in the nineties and your SAT/ACT are high, your chance for financial aid is increased. But other qualifications are equally as important. Leadership ability is highly valued, and extra-curricular activities that indicate your desire to contribute to the betterment of society is admirable. Therefore, your student resume is going to give a full and complete list of all your accomplishments, interests and activities. Be sure to include all your activities and interests both inside and outside of school. Remember, most scholarship committees are searching for students who make good grades, are involved in school activities, work part time, are involved in the community with their church or civic related projects, and demonstrate strong leadership and an interest in helping others in some way. So, if you are lacking in one of the areas that the scholarship committee may determine to be important, take care of that. If you have not done much to help your community, join something. There are all types of volunteer projects that need a student's help. Sign up to pick up trash or clean up the riverbanks when it's the local weekend to clean up the Trinity River. Or if you do not have many area activities, join an organization, such as, Friends of the Museum or the Symphony, and make a donation in your name and visit the museum or attend a symphony production. The local teen courts need volunteers. There are many organizations that appreciate volunteers. Being a volunteer helps everybody.

The following student resume and sample can serve as an idea for constructing your own personal resume. It is intended to provide you with ideas in order to prepare your student profile. It is only a guide.

Scholarship Resume Recommendation

Name
Address
Phone Number
Social Security Number

Personal Data

Date of birth	Weight (optional; i.e., athlete, gymnast, dancer, etc.)
Birthplace	Height (optional; i.e., athlete, gymnast, dancer, etc.)

Citizenship	Social Security Number
Marital Status	Number of children

Objective

State clearly and concisely why you want to go to college and give your professional or occupational preference.

Scholastic Record

Describe your scholastic record as fully as you can, including grades, standing in class, and awards and honors you have won.

Knowledge of Foreign Languages

State how many foreign languages you command. Indicate whether your knowledge is a result of lessons, or if you have learned them from your family, or if you have lived in another country.

College Entrance Requirements

Demonstrate that you can meet college entrance requirements by indicating how many units of each subject listed below you have taken. It is assumed that you will have checked previously to insure that you have taken all required subjects. Also, indicate your SAT and/or ACT score.

English	Trigonometry
Social Studies	Science
History	Chemistry
Algebra	Physics
Plane Geometry	Other Subjects

SAT ____________ and/or ACT Score ______________

Hobbies

List all hobbies and favorite interests.

Clubs and Associations

Give name and type of organization.
Name
Social Security Number

Extra-Curricular Activities

Describe all extra-curricular activities and the extent to which they have absorbed your time. Activities both within school and outside of school are important.

Employment Experience

Full information on employment is vital. State name of company, supervisor, duties and responsibilities, and length of time for each job.

Skills

List your skills. Include the various computer programs that you are familiar with.

References

Copy the most important paragraphs from each letter of reference you have in your possession. Retain the letters of recommendation to include in the complete packet or application.

Example of a Student Resume

Chandler Smith
123-45-6789
1633 Amber Lane
Arlington, TX 76011
817.555.1234
chs12345@msn.com

ACADEMIC HONORS
Who's Who Among American High School Students, 2000, 2001
Duke University Talent Search
National Honor Society
President's Award
Class Rank 35/550 Top 6%, Pebble Beach High School
3.9 GPA, SAT-Verbal 550, Math 650

FOREIGN LANGUAGE
Spanish, 2 years in high school, Conversational Spanish

HOBBIES
Golf, Baseball, Skiing, Camping, Music

EXTRA-CURRICULAR ACTIVITIES
Varsity Baseball State Playoffs
River Legacy Celebrity Golf Tournament
Student Council President

COMMUNITY/VOLUNTEER ACTIVITIES
Habitat for Humanity
New World United Methodist Youth Choir
Teen Court Volunteer
Baseball Summer Camp Volunteer

WORK EXPERIENCE
Summer, 2000, Parking lot attendant, Texas Rangers Baseball
2000-2001, Part time, cashier, Texas Motor Speedway
Summer, 2001, Lifeguard, Glen Lake Methodist Camp
2001-2002, Part time, Tutor math and science students

REFERENCES
Chemistry teacher: "Chandler is a mature and responsible young man."
Baseball coach: "Chandler works hard to achieve goals and is a leader."

TIMELINE

Junior Year of High School

Take the PSAT/NMSQT, SAT, ACT, and/or other Achievement Tests. Make sure your scores are sent to the colleges to which you may submit applications in the future.

Review your career plans and decide which type of college is right for you - large or small, public or private, two-year or four-year. Contact the admissions office of each college in which you are interested to obtain information on admissions, scholarships, and financial aid. Be sure to set up a file where you can keep track of all the information from each school. Include the names of the administrators that you meet, especially in the financial aid office.

Attend financial aid nights and college fairs to gather information on the colleges in which you are interested or check the Internet for college websites. Begin the college search process by reviewing as many guides and college catalogs as possible. Set up tours at your choices of colleges. Go to a game, walk around the campus, visit restaurants near the campus, and talk to students who attend the school.

Summer before Senior Year of High School

Create your own personal data bank of all the information that you may need to complete applications for colleges; scholarship applications, FAFSA, and any other form required pertaining to college.

Write several basic essays and be prepared to revise one at the last minute to meet a deadline for a scholarship application.

Visit the campus of several colleges and try to narrow the choices to three schools.

Create the winning student resume.

Begin your offline and online search for possible scholarships that match your interests, skills, academics, and talents. Create a list of scholarship possibilities from your online and offline searches to use. Be sure to watch the deadlines for each. Retain the contact information for future use, including name of organization, eligibility, prizes, what is required, and deadlines. If possible, download the application form or write for an official application at the proper time.

Remember to enclose a stamped self-addressed envelope. Use big, beautiful, bright, and unusual stamps on all your correspondence. Use correct postage.

Another very important step is to retain a copy of every item in the application package, including the essay, and the date it was mailed. You may need to review it for an interview or follow-up later to collect the award.

Fall Semester of Senior Year

Request letters of recommendation from: counselor, two teachers, employer, volunteer contact, close adult family friend, and pastor or adult friend in church. Provide each person a copy of your resume to inform them of your high school work, employment, and activities. Send them a note of thanks for the letter.

Visit your choices of colleges if needed and reduce your choices to two or three and apply for early decision admission. Also, find out how early you should apply for housing and financial aid. Some schools are crowded and do not have enough dorm space for all the students who request a room. Investigate the private dorms as a backup.

Sign up to retake the SAT/ACT to better your score. Order scores to be sent to possible colleges you will be attending next year.

Watch the mail for college acceptance letters and financial aid award letters to determine the college and package that best meets your needs.

Compare the financial aid award letters received and watch for response deadlines. You must sign and return a copy of your financial aid award letter to receive financial aid. Also, once you have selected a college, notify the colleges you decide not to attend, (only the ones for which you have received admission). Remember, someone may be waiting for that slot to open for him.

Spring Semester of Senior Year

As soon as possible after January 1st, complete the FAFSA and submit electronically or by mail. You will need student and parents' W2 form, most recent tax return, student driver's license number, and social security number.

Using the list of scholarships that you found online, offline, and from the counselor's office or PTA listing, begin requesting formal applications. If the application forms are available online, use them and start applying. Watch the deadlines. Notice whether or not the counselor or principals or teacher signature is required. Give them time to read and sign. Observe the rules, does it say postmarked by a certain date or received by a certain date? Does it say mail or turn in to the counselor? Pay attention to see if the application must be notarized. Beware of a postmark deadline being Sunday. Mail or submit early, but if you cannot, find out now if your city has a postal branch open twenty-four hours. It is not much fun to drive to the post office a few minutes before midnight to obtain a postmark to beat the deadline. But it is even worse to miss an opportunity to apply to a good scholarship because you missed the deadline. It is important to follow the eligibility rules. Watch for scholarship cutoffs that not only have a date, but also a time of day as part of the deadline. Read very carefully; it will make your scholarship search easier and more successful. Make sure that if the instructions specify original or official documents, you follow all requirements. If the instructions say copies are ok, use the copies and save originals for another scholarship.

Use your winning resume and all of the following items to create an impressive package for the scholarship committee to review. Unless the scholarship instructions strictly forbid any additional pieces of paper, then overwhelm them with all the information.

- Cover, thank you letter of introduction
- Student resume
- Application
- Transcript and explanation of GPA conversion scale
- Letters of recommendation
- Letter of acceptance from college
- Copy of anticipated college costs for one year
- Selective service registration for male students

Organization and early preparation will help you, as well as your parents, maintain sanity and friendship. All the miscellaneous information for filling out applications should be divided and kept in folders. Visit your local copy center and make copies of each letter, transcript, college costs, acceptance letter, tax return, selective service, and student resume. Be aware that some scholarship or college applications require the transcript or letter of recommendation to be in a sealed envelope. Some scholarship applications accept copies of transcripts and other materials while some organizations require originals. Read carefully.

Continue the scholarship search online and offline. Check your online search engines weekly to see if there are any new matches for you. Check with the counselor for any new scholarship notices that have been sent to your high school. Read the newspaper and magazines; keep your eyes open for articles on scholarships. Usually, the first scholarship deadline is in January. The busiest time is March, April, and May. There are more deadlines during these three months. Even though it is a very busy time for seniors with projects, finals, prom, banquets, ball games, etc., the best way to stay on track with your scholarship search is to prepare early, stay organized, and watch those deadlines.

Notify your college about any outside scholarships, grants or other funding you have been awarded. Ask your counselor to send final records to the college you have chosen. Finalize summer school or summer job plans.

Find out when your college is hosting its orientation session and plan to attend. It is a great way to meet classmates, faculty and staff, as well as get to know the campus.

Start making a list of things you will need for the college dorm. Work hard, and save your money for next fall. You will be a full-time college student.

Tips for Applications

1. Complete your application when you have adequate time to do so, in a quiet place without distractions. Take your time. Do a rough draft on a copy before starting the original.
2. Read it through and follow directions precisely.
3. Print neatly in black ink.
4. Use your real name, not nicknames.
5. Answer questions with concise, truthful answers.
6. Check your spelling and proofread the draft and original. Be sure to submit the original form.
7. If you make mistakes, use whiteout liquid and rewrite the response.
8. If an application fee is required, enclose a check or money order or provide credit card information with your application.
9. Keep a copy of your application, including essay and any other enclosures for your records. You may need to review it later at a scholarship foundation awards program.
10. Be sure to sign and date each application and include all of the required materials.
11. Mail your application early enough to meet deadlines. Be sure to use correct postage. Also, use the biggest, brightest commemorative stamps. Your envelope will get noticed.

Ten Most Common Errors on the Application for Admission

1. Applicants fail to meet the deadline.
2. SAT and/or ACT scores are not sent directly from the testing agency.
3. Unofficial transcripts are mailed with the application.
4. Official transcripts are not sent until after the deadline.
5. Transcripts are sent without a designated class rank.
6. Freshmen who rank in the top ten percent of their class do not complete the application and/or send in all required documents. This includes the required essays.
7. Questions on the application are left blank.
8. Failure to hit "submit" when applicants finish their electronic application.
9. Students use acronyms when describing extracurricular activities.
10. Students mark the wrong college code on their SAT/ACT tests.

Mistakes are bound to happen. These errors that reoccur among applicants are specific to the electronic application, while others occur on both the paper and electronic applications. These are common errors that students should avoid when applying to a university. Students should check the status of their applications via the Internet or contact the university well before the deadline. The earlier a student discovers that test scores have not been received or that transcripts never arrived, the better the chance of rectifying the situation before it's too late. Also, be sure to put correct postage if the applications are sent via mail.

Tips for the Interview

Before your application for scholarship assistance is granted, you may be asked to attend a personal interview. If it is possible, you will meet the scholarship committee or organization face to face. The interview is a formality; it is designed to enable the interviewer to form an impression of you. Therefore, you should give thought to the questions you will be asked and respond honestly. The scholarship committee will review your application and ask questions. They may ask your major, why you chose that college, will you be working or how many hours will you be working during the school year, what is your occupational or vocational goal, why you need scholarship assistance, or why you applied for help to this particular foundation. Be prepared to answer questions similar to these.

To create a favorable impression, do the following:

1. Arrive on time. It is better to be a few minutes early, than late.
2. Select the clothes you wear with care. Make sure your clothes are clean and pressed. Jeans would be out of place; business casual is more appropriate.
3. Be polite and friendly to the receptionist who greets you. She may be asked about your behavior.
4. Do not smoke or use any tobacco products. Do not chew gum.
5. When you are called in for your interview, remain standing until invited to sit down.
6. Sit up straight in the chair.
7. Try and relax. The interviewer is sincerely interested in helping you.
8. Make your answers enthusiastic and informative, and keep to the point.
9. Be truthful.
10. When the interview is finished, stand up, thank the interviewer for his courtesy in seeing you and leave as gracefully as possible.

The Follow-up Letter after the Interview

After the interview, same day if possible, write a short note thanking the interviewer or the committee for having seen you. It will help create the impression of you as a considerate and worthy candidate. If you have promised to send any additional information, attach it to your letter. Sometime after the interview, you will be notified that you have been accepted or rejected as a scholarship student. If you are accepted, you will want to respond and express your thanks. You may ask what the procedure is to collect the scholarship. Is the check sent to the college, or is it sent directly to the student, or is there any college student verification needed? There are many ways for a scholarship to be awarded. If you were rejected by the scholarship committee, you want to keep your name before the board and ask if you may apply for financial help at a later date. They will look at you with favor if you prove yourself able to accept defeat gracefully. They will also appreciate a reply of thanks. Without their generosity, fewer students could attend college.

Example of Follow-Up Letter after the Interview

Date

Name of Interviewer
Name of College or organization
Address
City, State Zip

Dear _________:

I want to express my sincere thanks to you and the entire committee for having granted me a personal interview. It was an opportunity I will always appreciate.

I am eagerly waiting for your decision in regard to my application for acceptance (or scholarship assistance), hoping that it will enable me to begin my academic career at _____________ (name of college).

Sincerely,

Your name
Your address
Your phone
Your email

Do not overlook the opportunity to thank the generous foundation, college, business, or association for the privilege to be a recipient of their prestigious scholarship. Following is an example of a thank you letter.

Example of Letter After Notice of Acceptance

Date

Name of Interviewer
Name of college or organization
Address
City, State Zip

Dear ______________:

I have just received the exciting news that I have been awarded the Arboretum Academic Scholarship in the amount of $1,000 per year. I want to thank you and the committee for your favorable consideration. As I begin my ____________ (freshman, sophomore, junior, senior) year at ___________ (your college), I want to say thank you for helping my family and me.

Sincerely,

Your name
Your address
Your phone
Your email

4 How To Beat the Competitors for an Athletic Scholarship

A higher education combined with a competitive athletic program will undoubtedly be a uniquely rewarding experience. Becoming a college athlete is not simple, and it will take hard work and a lot of cooperation from parents and the coach. If you are not a superstar, you will need to market yourself to colleges to maximize chances of an athletic scholarship.

Sports recruitment is a highly sophisticated process that usually must be initiated by athletes themselves, with the exceptions of the extraordinary athletes who participate in high profile sports. Often, coaches miss an opportunity to meet a student who could help his program simply because they do not know that the athlete exists. Only one to two percent of college-bound students qualify for athletic awards. This low percentage is a result of guidelines established by the National Collegiate Athletic Association (NCAA) that require each sport to award a specified number of scholarships for each year or competitive season. Therefore, athletic aid may not be available to first-year students if all awards have been previously assigned.

Many college athletic programs are regulated by the National Collegiate Athletic Association, which is an organization founded in 1906 that established rules on eligibility, recruiting, and financial aid. The NCAA has three membership divisions -Division I, Division II, and Division III. Institutions are members of one or another division according to the size and scope of their athletic programs and whether they provide athletic scholarships. If you are planning to enroll in college as a freshman and you wish to participate in Division I or Division II athletics, you must be certified by the NCAA Initial-Eligibility Clearinghouse. The Clearinghouse was established by the NCAA member institutions in January 1993. The Clearinghouse ensures consistent application of NCAA initial-eligibility requirements for all prospective student athletes at all member institutions. It is the student's responsibility to make sure the Clearinghouse has the documents it needs to certify you. The documents are: your completed and signed student release form and fee, foreign student application (if applicable), your official transcript mailed directly from every high school you have attended, and your ACT or SAT scores. The publication describing the NCAA and procedures is available from your high school counselor, check the internet, or write for a brochure at NCAA Clearinghouse, P.O. Box 4044, Iowa City, IA 52243-4044.

Even more important are the athletic and academic eligibility requirements you must meet in order to compete at the collegiate level. Also, where you train has a direct impact on your visibility and recruitment. In general, athletes train in private clubs or in high school programs. Club sports have the highest rate of recruitment, which could include diving, fencing, golf, gymnastics, lacrosse, rifle, skiing, soccer, softball, swimming, tennis, water polo and volleyball. High school sports that have the highest rate of recruitment are baseball, basketball, cross-country, field hockey, football, ice hockey, lacrosse, track and field, and wrestling. These are all recognized NCAA sports. However, there are several emerging sports that the NCAA seeks to promote which include archery, badminton, bowling, crew, squash, synchronized swimming, and team handball.

Athletic departments of colleges and universities sponsor most athletic scholarships. The biggest schools will persuade the best athletes in the most popular sports. If you can not compete in that arena, look at smaller colleges and universities with ambitious sports programs. Though your skills may not have gotten you far at State U, the smaller college may be happy to have you and willing to provide scholarship money to entice you to attend.

If you decide to pursue athletic scholarships, start searching for them early, as early as your sophomore year in high school. Your coach can probably recommend colleges that have strong programs in your sport. Numerous college guides break down athletic scholarships by sport to make your search easier. Contact the coaches for your sport at each college you are interested in attending, and inquire about scholarship opportunities. Put together an athletic resume that lists your accomplishments in the sport, your stats, your coach's name, the camps or summer programs you have attended, the honors you have won, and anything else that you think will improve your chances for winning a scholarship. You might also include a letter of recommendation from your coach, newspaper clippings about your athletic accomplishments, a video showing you in action, and of course, an excellent student resume. A good GPA will impress the coaches too.

The main goal in the recruitment process is to make yourself known to college coaches. The following action will help you do that.

1. Apply to colleges that compete on a dimensional level at which you are capable of playing.
2. Match your preferences with the college of your choice. Choose the school that will best utilize your athletic talents.
3. Once you have selected a college or colleges, contact, by phone and by mail, the coach of the sport you are interested in playing.
4. Send newspaper clippings to the coach.
5. Send a video of yourself in action.
6. Mail a copy of your competitive schedule.
7. Be able to communicate your athletic goals to the college coach.
8. Get letters of recommendation from your high school or club coach, your counselor, adult friends, teachers, trainer, and employers.
9. Plan to attend a summer sports camp during your junior or senior year.
10. Be prepared to answer the following question. What do you want to do besides play sports? This refers to your academic choices, community involvement, and career plans.
11. Visit the campus and coach and tour the athletic facility. Generally, college coaches select students somewhere between the summer following the athlete's junior year or the spring of his or her senior year in high school. When a coach decides whether or not to seek a student, he or she is essentially determining if the athlete will fit into their particular program.

Some popular traits coaches look for in athletes are that they are serious about education, motivated, dedicated, respectful, hard-working, disciplined, determined to excel, outgoing, always ready for improvement, team-oriented, and does not rest on past accomplishments. Review the sample letter to a coach.

By following these suggestions for searching for an athletic scholarship, your chances will be maximized, especially when sending a winning student profile along with your highlighted list of sports accomplishments, awards, and honors. Remember to be both persistent and assertive in gaining a position on the college sports team of your choice.

Example Letter to Coach

Date

Name of Coach
Name of College
Address of College
City, State Zip

Dear Coach ___________,

I am presently attending ________high school and will be graduating with the class of _______. I am very interested in continuing to compete at the college level and would like to find out what your _____(baseball, football, volleyball, etc.) program has to offer.

Please send me catalogs, schedule or other literature regarding ______ (name of the college) and team information.

Enclosed is my resume. Please review my qualifications and interests to see if they would match your school. Let me know if you need any further information.

I look forward to hearing from you.

Sincerely,

Your name
Your address
Your phone
Your email

Enclosure

5 How to Create Extraordinary Essays

There is no way to avoid writing at least one essay. You can save yourself some grief by starting to think about essay topics and jotting down ideas early. How important is the essay? Very important. It is your opportunity to take charge of the impression you make on a college or a scholarship committee and to provide information that does not appear in grades, test scores, or other materials. The essay allows you to reveal your talent, intelligence, sense of humor, enthusiasm, maturity, creativity, sincerity, and writing ability; they all count in an admission evaluation.

The admissions people and scholarship judges read millions of really boring essays about "how wonderful I am" and "my plan to end world hunger." Colleges are very worried that their students do not write as well as students did in the past. If you have strong writing skills, you have a serious edge. When admissions officers read essays, they search for evidence of curiosity, strong moral character and the capacity to commit to meaningful endeavors. Admissions personnel want to get a sense of your unique voice and distinctive qualities. Write in a way that holds the reader's interest. Make yourself come alive on paper.

You want to get your point across, not bury it in words. Your prose should be clear and direct. If an admissions officer has to struggle to figure out what you are trying to say, you will be in trouble. Avoid adjectives and exclamation points. As you reread your essays, stop at every adjective and ask yourself if it is necessary. Do not bother asking about exclamation points - leave them out.

If your essay is filled with misspellings and grammatical errors, admissions representatives will conclude not only that you do not know how to write but also that you are not smart enough to get help. Ask your mom, dad, teacher, brother, sister, someone, to read your essay and comment on it. Use spell check. Be especially careful about punctuation.

Here are a few recommendations for developing essays that present your personal qualities.

1. Do not put off writing your essays until the last minute. Plan them during the summer before your senior year, if you can, or early in your senior year. Give yourself time for all the steps below, and write an individual essay for each college.
2. Be sure you understand the college's topics, directions, and deadlines.
3. Before you start your essay, list your aspirations and how you think the college will help you meet them. Review your list of school and extracurricular activities, travels, and awards or other accomplishments. Include work experiences. List any academic or personal weaknesses you are trying to improve, as well as the traits you value about yourself. To focus your essay, study the lists and develop a one-sentence theme from it.
4. Write a rough draft. Set it aside for 24 hours. Then read to check for any grammar errors, rambling, or boring lines. Rewrite your essay based on this evaluation and sharpen the essay.
5. Ask someone whose opinions you respect to read your essay and give his or her impression. Ask for specifics. What do you think I am trying to say? How do I come across as a person? What parts confuse or bore you? Where do I need more details? Which parts did you like best? Rewrite your essay and smooth the rough spots. Read the essay aloud. This will help you spot questionable areas.
6. Mail your essay on time and relax. If you have done all of the above, you can be sure your work will be read with appreciation.

Here are some common essay topics or questions, followed by typical responses to them and ways to avoid getting stuck in that rut.

What have you chosen as your major and why?

Why should you be selected for the scholarship and how will society benefit?

Relate an incident from your life that reveals a significant aspect of your character.

Write about something you have read. Many applicants think this means you have to do a book report. But that is the danger: Too many responses to this sort of question sound like book reports, and nobody likes to read book reports. Do not write about an obvious book. Your choice should make you seem interesting. Write about someone you admire. Most people write without much conviction about the President, a relative, or Martin Luther King, Jr. These people are highly admirable and heroic, but they seldom inspire interesting essays. Instead, find someone truly meaningful to you. Your hero does not have to be famous, you can choose your favorite uncle who took you to your first collegiate football game. That day made such an impression, you haven't thought of attending any other university. Finding someone truly meaningful is better and your choice will surprise the reader.

Why do you want to attend this school? Your real reason may be something better left unsaid ("Because I want to party for four years"), but the closer to the truth your answer is, the more likely you'll write something enjoyable. Avoid generalities ("to get a good liberal arts education," "to broaden my knowledge"); be as specific as possible.

How do you see yourself ten years from now? Avoid describing your perfect job. This is a tough question because the temptation to be boring can be overpowering. Narrow your focus. Picture yourself doing something a bit unusual, and do not strain to make a direct connection between it and a college education. Do not pretend that you think your college degree will be the sole foundation on which you build the rest of your life. Write about a meaningful activity. Which of your activities really was the most meaningful? Which was the most fun? Don't say that your most meaningful activity was your service on the Student Council because it taught you the importance of effective leadership. The essay judges hear that often. Instead, discuss the qualities you developed while you coordinated special programs and volunteer activities and managed to maintain your GPA. List the fun stuff from being on Student Council, how many people you met at the state convention, which new ideas for student involvement will you introduce to the school. Create an interesting paper for the committee to read. It will make the difference.

6 How to Benefit from a Campus Visit

College admissions officers report that the first rush of campus visits occurs as juniors begin spring break. Summer is fine too. What is important is that you and your parents visit. Videos, brochures, and Internet images are good starting points, but those who skip the actual visit increase the likelihood of a poor match. Admissions officials say that poor matches contribute significantly to the nation's 27 percent dropout rate among college freshmen. By fall of your senior year in high school, you should narrow the choices to three to five schools, which include a dream school, if desired, and an academically and financially safe school.

Now, for many high school juniors and seniors, touring college campuses is as common a ritual as getting a driver's license or going to the prom. Summer has become the prime season for campus visits, as many families now plan summer vacations around these tours, dragging along bored younger siblings who can only count the hours until they can hit the hotel pool. When taking campus tours, have some fun. Alternate campus tours with afternoons at the beach or a theme park. Even better, parents can split up so that the rest of the family doesn't have to spend the whole day on campus.

More than 60 percent of this year's seniors will head to college, up from 50 percent in 1980. As a result of the increased competition, the desire to get the most bangs for the buck, and the advent of the Internet, high school seniors are visiting and applying to more schools. It is no longer uncommon for families to spend thousands of dollars on cross-country trips to campuses before their children have applied to college, let alone been admitted.

Do your homework before you arrive at each campus. Take several college guides in the car and a folder on each school so that you can read about each place as you drive into town. Make sure you have a chance to talk to other students on campus and get a sense of the community. If you are brave, arrange to have dinner on campus or stay overnight in a dorm. Even if you don't take notes during the visit, write your thoughts down while the experience is still fresh in your mind.

Selecting the best school for you is as individual as finding a good fitting shoe. You have to try it on, walk around, and feel like you are getting a good value. How can you do that with a college? The answer is easy. Visit the campus, talk to students and staff, and compare schools. If you have a particular area of interest in which you want to pursue your degree, ask to meet with a professor in your field of interest. Also, check out access to computers, campus security issues, medical facilities, nearest mall, movie theater, restaurants, churches, and public transportation. Ask the following questions while on campus.

When you talk to current students, ask:

1. How easy is it to get registered for the classes I need? Are some hard to get into?
2. Are faculty members interested in their students and accessible outside of class?
3. Is the college library a good place to study and do research?
4. Are campus jobs readily available?

As you tour the campus, ask yourself…

1. Are the rooms in the residence halls pleasant and quiet enough for studying?
2. What's the cafeteria like?
3. What's the surrounding town or city like?
4. Are there activities that interest me on campus or in the surrounding community?

When you talk to an admissions representative, ask…

1. Does the college have academic programs that fit my interests?
2. Will I have access to computers and other special equipment?
3. What extra-curricular activities are there on campus and what are the facilities like?

4. What is the job placement rate for recent graduates in my field of study?
5. What new programs or facilities will be available in the near future?
6. What financial aid assistance is available from the school?

When your visit is over, ask yourself...

1. How do I feel about the academic demands and atmosphere?
2. Were the students the kind of people I'd like to get to know?
3. Did I like the social atmosphere?
4. Am I comfortable with the size of the college and the classes?
5. Do I feel comfortable and safe on campus?
6. Is the weather ok for me? Some like warmer climates, some students enjoy colder weather. What kind of temperatures do you like?

Remember, you are paying for a quality education. Make sure you get it. It is your choice.

7

How to Avoid the Scholarship Scams

Need money for college? Doesn't everybody? Families searching for financial aid are targets for fraud. Unfortunately, in their efforts to pay the bills, many students and their families are becoming victims of scholarship scams. Be skeptical and cautious if you hear one of the following promises.

- The scholarship is guaranteed or your money back.
- You cannot get this information anywhere else.
- I just need your credit card or bank account number to hold this scholarship.
- We will do all the work.
- The scholarship will cost some money.
- You have been selected by a national foundation to receive a scholarship.
- You are a finalist in a contest you never entered.
- This is your last opportunity and you are a fool if you don't take it.
- Give me the names and addresses of your friends who have senior students so I can help them.

The sad truth is there are scammers out there posing as scholarship providers or scholarship matching services that take your money and leave you with nothing. The good news is you can protect yourself from scholarship scams by following these tips.

The phony scholarship: You receive a notice about a scholarship that promises you cash. All you need to do is pay the registration fee. So why not apply? Because there is no award. Or there is one small award used to lure you into the trap.

The phony scholarship matching service: Pay a fee, and they will do all the work. They will find information that you can't get anywhere else. Not only that, they will guarantee that you'll win. The outcome? You never hear from them again or you receive a list of inappropriate or obsolete awards. And forget your money-back guarantee: The company has disappeared.

The phony educational loan: a low-interest loan in exchange for an up-front fee. You pay the fee, but never receive the loan.

The phony free financial-aid seminar: a thinly disguised sales pitch for a bogus scholarship search or insurance offer.

A legitimate scholarship provider sends information about awards only after you request it.

A legitimate scholarship organization makes its contact information available, including a telephone number and address, as well as references.

A legitimate scholarship matching company never guarantees that the student will win an award, and they never promise to "do all the work" for the student.

A legitimate scholarship application requires only information that is relevant to the award. A legitimate educational loan company deducts fees from your disbursement checks; they don't collect the money up front.

Remember, don't be fooled by an "official sounding" name. Just because a company uses words like National, Federal, Foundation, or Administration in its title doesn't mean it is a legitimate operation.

This is what to do if you suspect a scam. Document all your dealings with any company that you suspect of fraud. Include details about the offer, your response and the dates of your communications. Take notes during any telephone conversations with these organizations. Record the date and time of the conversation, the name of the person with whom you spoke and a detailed account of your conversation. Report them. Any of the following organizations can help:

National Fraud Information Center (NFIC) 1-800-876-7060
www.fraud.org or write:
National Fraud Information Center
POB 65868
Washington, DC 20035

Federal Trade Commission (FTC), 1-888-FTC-HELP (1-888-382-4357)
www.ftc.gov or write:
Federal Trade Commission, CRC-240
600 Pennsylvania Avenue, NW
Washington, DC 20580

Better Business Bureau (BBB) 1-703-276-0100
www.bbb.org or write:
Council of Better Business Bureaus
4200 Wilson Blvd., #800
Arlington, VA 22203-1838

US Postal Inspection Service (USPIS) 1-800-654-8896
www.usps.gov or write:
United States Postal Service, Attention: Hotline
1735 N. Lynn Street
Arlington, VA 22209-2020

8 How to Find a Low or Zero Interest Loan – Financial Websites

A loan is a form of financial aid that must be repaid with interest. Scholarships and grants do not have to be repaid. Your main loan options are:

- Student Loans (also called Stafford and Perkins Loans)
- Parent Loans (also called PLUS)
- Private Loans (also called Alternative)
- Loan Consolidation

Few students can afford to pay for college without some form of education financing. The typical undergraduate student completes college with $16,500 in student loans. Graduate and professional students borrow even more.

Grants, scholarships, work-study and other forms of gift aid just do not cover the full cost of a college education. Many students find that they must supplement their savings with government and private loans. The Federal education loan programs offer lower interest rates and more flexible repayment plans than most consumer loans, making them an attractive way to finance your education.

Education loans come in three major categories: student loans, parent loans, and private loans. A fourth type, consolidation loans, allows the borrower to combine all of their loans into one loan for simplified payment.

Many student loan providers offer low cost government and private loans with high-quality servicing and flexible repayment terms. You may apply online for your loans at various banks or use the Lender Codes Database to obtain the lender codes of other lenders.

Loan forgiveness programs, in which the borrower's loans are paid off in exchange for volunteer work or military service, offer an option for easy repayment. If you are having difficulty repaying your education loans, see Defaulting on Student Loans before you decide to skip a payment. It offers you some alternatives.

Also, FinAid provides numerous calculators that can help you better understand your borrowing options. The loan calculators offer estimates of monthly loan payments, estimates of the amount of debt you can afford to repay, an analysis of the cost of capitalizing the interest and tools for comparing loan costs.

If you need a loan to help pay for the cost of college, you are in good company. About half of the class of 1998 will incur student-loan debt at an average of $13,600 per student, resulting in monthly payments of $163 for the next 10 years, according to the Student Loan Marketing Association. Sobering news? Perhaps, but student loans are a good deal. They are granted at below-market interest rates, require no collateral, and repayment typically does not begin until after graduation. While your loan may come directly from colleges, state governments or private lenders, the federal government is the largest provider. Start your search with an overview. Visit the National Education Loan Center and Crestar Student Lending for additional information. They'll walk you through some of the more typical loan programs.

You'll find that financial need dictates whether you'll qualify for favorable interest rates under a Perkins loan or for a delayed repayment schedule under a Stafford loan. Standards are generally much the same as they are for student aid.

Perkins Loans

The Perkins loan program requires a high degree of financial need and is available to graduates and undergraduates. The program is administered through your college, which is the actual lender. Eligibility is determined on the same basis as student aid. To apply, file the Free Application for Federal Student Aid (FAFSA) form with your financial aid office.

A Perkins loan is one of the most advantageous of all loans. Interest rates are only 5%. There are no origination fees. Payments don't begin until nine months after graduation or leaving school, and usually extend over a 10-year period. The loan is canceled if you serve in Vista or the Peace Corps.

The yearly maximum loan under a Perkins is $3,000 for undergraduates and $5,000 for graduates, with a lifetime maximum of

$15,000 and $30,000, respectively. But if you attend a college with a good payback rating, the maximum loan can be increased. When choosing a college, don't hesitate to ask the admissions office about past payment rates. Explain that you believe you're eligible for a Perkins, and the overall record of the institution can affect the size of your loan.

Stafford Loans

Unlike the Perkins loan, the Stafford is available to all students. It accounts for more than half of all student loan programs. For those demonstrating need, payments don't begin until six months after a student leaves school, and no interest accrues until that time. With an unsubsidized loan, interest accrues immediately after the funds are distributed, usually each semester.

Apply as you would for financial aid, by filing a FAFSA form directly with your college. The government runs the online Federal Student Aid Information Center. If you get an unsubsidized Stafford, you can elect to start payments six months after you leave school, but you must pay the interest that has accumulated throughout your school years.

Stafford rates vary and are adjusted each July, but are capped at 8.25%. As of July 1, 1997, the rate was 7.66%. Starting in July 1998, the rate will be tied to the 10-year Treasury bill. In addition to interest, there is an origination fee of 4% to 5% of the total loan amortized over the life of the loan.

Loan limits are $2,625 for the first year of undergraduate school, increasing yearly to $8,500 for graduate and professional school. There is a lifetime cap of $23,000 and $65,500, respectively.

Prior to obtaining a Stafford loan, you must participate in a loan counseling session that drives home the importance of debt management. A Website sponsored by various banks allows you to take the session online and will send your "quiz" results directly to participating schools. If your school is not among the several hundred that participate, you can still use the site for practice.

Once you get the loan, the U.S. Department of Education will make payments to you through your school, usually in two payments a year.

Private Stafford Loans

If your school does not participate in the Federal Direct Student Loan Program, banks, credit unions and private lenders offer unsubsidized Staffords. Unfortunately, you may not have as many ways to repay. But if you pay your bills on time, the student loan consolidation group, Sallie Mae, may buy your loan from the private lender. If your loan is bought and the first four years of payments are made on time, the interest rate is lowered by 2 percentage points on the remaining loan balance. You can save another 0.25% by authorizing automatic deductions from your bank account. If there are two years of on-time payments, Sallie Mae "forgives" the origination fees of more than $250. These steps will allow you to save $386 on a $5,000 loan and up to $7,095 on a $60,000 loan.

PLUS Loans

PLUS loans, or Parent's Loan to Undergraduate Students, are issued to parents by banks and special lenders. These loans are designed for parents with good credit histories whose dependent children attend school at least half time. PLUS loans are not needs based, and the loan limit equals the cost of attendance minus any aid the student receives. Parents may apply directly to the school or to an independent lender or state guarantee agency.

Interest rates are currently about 9%, and are adjusted each July 1. They track the one-year Treasury bond, with another 3.1 percentage points added, capped at a maximum of 9%. There is a 4% origination fee. Payments begin 60 days after the last loan disbursement for the academic year, so both interest and principal is paid while at school. The loans are repaid over a 10-year period.

State Loan Programs

State-guaranteed student loans are available from almost every state, with interest rates of about 8% for undergraduates and 12% for graduate students. The average state loan is $2,500 for undergraduates and $5,000 for graduates. Apply to banks that administer the state programs. The individual college's financial aid officer will make the ultimate decision.

Small and Short-Term Loan Programs

The College Board Extra Credit Loan imposes interest based on the 90-day Treasury note, plus 4.5 percentage points with a 15-year payback term. These loans are applied for on a yearly basis and are best used to help you if your aid is withdrawn or reduced.

The Complete Source for Financing Education and some banks offer small loans of $500 to $2,500, with 15-year terms at 5% interest.

Academic Management Services, which is affiliated with 1,500 schools, will foot the tuition bill as long as you pay them back within 10 months. College Resource Center will lend as much as $50,000, but you must begin to pay 2% of the outstanding balance per month, in the first month after the loan is approved. There is a 3% origination fee and interest comes in at 2.5% above prime.

Military Loans

The military offers a great way to save on college. For example, the child or spouse of a veteran who died or was seriously disabled because of military service may be entitled to an education benefit.

Government Loans

Go to the U.S. Department of Educations' Student Financial Assistance site. This is the federal government's main information center for financial aid for students. Here you'll find definitive information about federal aid programs, including the Pell Grants, Federal Direct Loans, Federal Family Education Loans, (FFEL), Federal Supplemental Educational Opportunity Grants, FSEOG), Federal Work-Study (FWS), and Federal Perkins Loans.

Students and parents often find that they do not get as much money through grants and scholarships as they need. If you find yourself in this situation, you may want to consider a loan. Lending institutions provide all types of loans. There are other loan options available if your budget needs a little increase in the cash flow area. Many private foundations offer no interest or low interest loans that are scheduled for repayment after graduation. A list of helpful websites for financial aid and loans are included in this section. Remember, college is an investment in yourself. And borrowing now will pay off later, because a college degree will increase your earning power. Once you decide to borrow, borrow only what you need, because usually you must pay back what you borrow, with interest. Estimate the amount of debt you may be able to afford and your monthly student loan payment before you borrow.

If a student loan is your first borrowing experience, consider this responsibility seriously; your ability to borrow in the future depends on it. Make sure you read and understand the terms and conditions on your promissory note. You are agreeing to repay the loan with all accrued and capitalized interest and deducted fees. You are obligated to repay your loan regardless of whether you complete your education, are satisfied with your education, or able to find a job. It is also your responsibility to read and keep all your records and to contact your lender regarding any changes in your status as a student. The amount you borrow and how conscientiously you pay it back will affect your ability to borrow for a car, house, or other purchases after graduation. Be aware that some prospective employers request a credit report before they hire new employees. If you are late with your student loan payments, it will have a negative effect on your credit history. The good news is that repaying your student loan on time can help you establish and maintain an excellent credit history.

If you fail to repay your loan according to the terms disclosed on your promissory note, you are considered to be in default. The consequences are serious. The delinquency and default will be reported to at least one national credit bureau and your credit history may be damaged for up to seven years. The entire amount of the loan, including interest, may be due immediately. Your wages may be withheld and your possessions may be seized. You may not be eligible for revised payment schedules or additional deferments. The loan may be referred to a collection agency and you may be liable for collection costs. Your federal and state tax refunds may be withheld.

If you experience repayment troubles, your lender can help. If you are temporarily unemployed, disabled or simply struggling, many lenders will work with you. You can be assured that most lenders will try to help you in one of the following ways. For example, a deferment allows you to postpone payments for reasons stated above and more. Forbearance is an agreement with your lender to postpone payments when you are not eligible for deferment. Even if you're employed, changing your payment schedule or consolidating your loans may help simplify your life and reduce your monthly payments.

The experience you gain as a student loan borrower and credit consumer can provide you with the financial foundation to meet the challenges of borrowing and using credit in the years ahead. Maintaining good grades and applying early are two significant strategies to increase your odds for receiving the best loan possible. Learning to budget, and making wise decisions about borrowing and using credit, are important steps toward managing your expenses during your college years and beyond.

Contacts for Loan Information

Congressional Hispanic Cansus Institute
1-800-EXCEL-DC
http://www.chci.org

Loan Star Lenders
1-800-829-4599
http://www.loanstar.com

Texas Guaranteed Student Loan Corp.
1-800-252-9743
http://www.tgslc.org

Texas Higher Education Coordinating Board
(Admission and Scholarship Info)
1-800-242-3062
http://www.thecb.state.tx.us

The Financial Aid Information Page
http://www.finaid.org

Hispanic Scholarship Foundation
1-415-892-9971
http://www.nhsf.org

Minnie Stevens Piper Foundation

http://www.window.state.tx.us/scholars/mspmain.html

www.atheu.com

http://lwww.adventuresineducation.org/sbase/

www.asapubt.com

www.bankamerica.com/studentunion

www.chase.com/educationfirst

http://www.castpress.com/

www.colleges-schools.org

U.S. Department of Education-/Student Financial Assistance

www.educaid.com

http://fdncenter.org

www.firstedcu.org

www.franklinlindsay.org

http://www.free-4u.com

http://iwcc.pair.com/scholarshipage/subjects/

www.loanstar.com

www.plato.org

www.millhollon.com

Abe and Annie Seibel Foundation, Galveston, TX

Hattie M. Strong Foundation, Washington, DC

www.salliemae.com

studentloan.com

http://www.tgslc.org/

www.usbank.com

www.unipac

Financial Aid Websites and Contacts

www.bestkeptsecretsforwinning.com

Ameri Corps

CollegeLink.com

www.collegeEdge.com

www.collegenet.com/

The College Board

www.chase.com/educationfirst

CollegeClub.com

www.collegeispossible.org

College Board Financial Aid Services

www.cns.go

cool-College Opportunity On Line

www.cns.go

embark.com

U.S. Department of Education

www.ed.gov.pubs

http://www.estudentloan.com

www.ed.gov.offices/OPE/

Estrong.com:College

easi.ed.gov

http://www.fenbd.com

FinAid.com

4Scholarships.com

FAFSA

Federal Pell Grants

www.fafsa.ed.gov

www.fastweb.com

First Bank Scholarships for Students

fastaid@fastap.org

www.finaid.org

www.finaid.org/calculators

www.ftc.gov

Freschinofo.com

www.goarmy.com

www.honoring.com

http://iwc.pair.com/scholarshipage/links.html

http://www.jayi.com/jay:

www.kaptest.com

www.kaploan.com

Money Central Family and College

National Student Loan Clearinghouse

http://www.naas.org

www.nasfaa.org

NASFA.org.est.efc

www.nelliemae.org

Octameron Associates

www.phecfld

Patterson's College Quest

www.rams.com/srn/

www.rspfunding.com

Scholarship Search Wizard (Fastweb)

Supercollege.com

The Scholarship Page

scholar@ecsf.org

Scholarships.com

StudentLoan.com

http://subscription.collegeview.com

www.salliemae.com

www.sourcepath.com/aid/

www.students.gov

web.studentservices.com

www.scripps.com/foundation

www.srnexpress.com

www.targetusa.com

Texas Tomorrow Fund

Theoldschool.org

www.wiredscholar.com

World Education Access

9

How to Manage Time and Stretch a Dollar

On your mark, get set, start the budget. College can be a great money management experience for both student and parents. It may be the first time that you are truly on your own and handling money without parental supervision. But, how much money will be needed? How can you economize and develop a spending style that keeps you and your parents worry free for the next four years?

You can start by recognizing that college costs, like many types of expenses, fall into two basic categories: fixed and flexible. Try to pre-pay the fixed expenses. Then develop the right spending habits to manage what is flexible. If expenses are well thought out, students won't have to take a loan from the Bank of Mom and Dad. For most of our adult lives we manage inflow and outflow of cash, and it's tough. Only in the area of college spending do students have the opportunity to concentrate on their spending habits.

If you are a student, take a few minutes to mentally walk through your college day. Will you use the meal plan for breakfast? Will you buy snacks? Will free periods involve any expenses? What are your typical weekend plans? What might they cost?

Here are a few of the non-tuition expenses that you should keep in mind as you prepare for college this year. By considering all of the potential costs, you can design a reasonable budget. The best budget is one that is discussed and prepared by the student and parents.

- Travel expenses to home, friend's houses or for a spring or Christmas vacation
- Entertainment
- Food
- Clothes

- Sports Equipment
- Phone calls home or anywhere
- Postage
- Haircuts
- Periodicals
- Internet service providers or other online access
- Medical and dental fees
- Insurance
- Miscellaneous needs
- Transportation costs

Students and parents should be realistic about social spending needs. You can do this by asking yourself to review the following financial exercises to learn about your spending habits. While these are suggested for students, parents would benefit from doing them in their own lives.

1. Keep a journal. Record your money habits for a week. Write down every penny you spend. See how much goes to newspapers or magazines you never read, candy you never eat and cigarettes you shouldn't smoke. How much goes for meals out when you have a pre-paid student meal plan? How much money goes for convenience, such as cab rides, dry cleaners, or take-out food? As a student, don't pre-judge your actions. Simply record what gave you satisfaction and what was a waste of money.
2. Create a chart of your spending habits. Make a budget expressed in percentages, not dollars. If you have $150 a week to spend and $50 is spent on food, that's one-third of your pie chart. By visually calculating your spending, it typically has a profound psychological effect and leads to more savings with less frustration than counting pennies. If, for example, your car takes up half your pie chart, you can begin to shift expenses away from other things to meet your auto expenses, or limit your car use if you conclude that it's not worth it. The bottom line is that spending is discretionary. It's up to you to allocate and do without if you are unrealistic.

You can also take an active stance to reduce unneeded expenses and to avoid the usual student/parent money fights. Here are some tips to avoid financial disagreements.

Instead of using a credit card, establish a debit card. Debit cards are limited by the amount of money held in the account of the bank that issued it.

Use a phone card or your cell phone minutes that cover what you expect to be reasonable costs over a given period of time, such as a month.

Never make a final budget until after the first two months of college. You can't really judge needs completely. Dashed expectations lead to disagreements. Estimate and stay flexible.

Be realistic about using a meal plan. Many students under-use them. See if the college will let you sign up a few weeks after school starts, instead of earlier, as they usually request.

In your enthusiasm, don't buy all new clothes for college. Take the essentials and ship or deliver what is missing, after it's been determined what's needed. This can leave more money for important things.

Consider online banking. Many banks now offer online banking capability, which offers much faster money transfers and also allows you to more quickly monitor cash outflow.

Do not expect freshman budgets to look like senior budgets. Things change with inflation, needs, and job schedules. If at all possible, try to find a job that will enhance your future career. A volunteer summer job at the state aquarium will get the future oceanographer further in the long run than working in a restaurant, day camp, or the mall.

10 How to Survive Freshman Year

College is a challenge for most students. You'll have more freedom than you've ever had. You may choose how many courses to take, what time to schedule them, even whether to attend every class or not. With this freedom comes a great deal of responsibility. If you goof off, sleep or party too much, you'll soon get behind in your studies and pay the price one way or another, by getting bad grades, failing, or having to drop classes and pay to take them again the next semester.

Most colleges offer classes or assistance in developing college skills such as note taking, studying, researching and writing, getting organized and managing time. Many also offer tutoring to help students master certain subjects, or they offer mentoring programs, which match a student with a faculty or staff member who can provide advice and guidance. Ask for help at the first sign of trouble. The longer you wait, the harder it will be to get back on track.

Keys to success in college:

Go to class. Students who attend class rarely fail.

Work hard. Work harder than you did in high school. Spend two to three hours out of class studying for every hour of the week you spend in class. For a fifteen hour class schedule, spend 30 hours a week studying and 15 in class.

Play hard. College is not meant to be all work. Relax and have some kind of balance in your life.

Participation. Participate in all the orientation activities that your college offers. Students who participate in these elective courses have a higher probability of becoming sophomores.

Many universities offer a course called "Success for College" and teach basic attributes of successful students, such as time and money management, extracurricular activities, class attendance, and degree planning. *Sign up and learn.*

Live on campus. Students who live on campus have a higher survival rate, especially during your first two years of college.

Choose your friends wisely. You are likely to become just like them. Winners select winners. Losers select losers.

Get help early. Do not wait until late in the semester. There is no stigma to seeking assistance. Remember, you're paying the salaries of the college employees and you have a right to ask them for help.

Develop a relationship with your academic advisor or counselor. If you don't have one initially who seems to show an interest in you and whom you feel you can relate to comfortably, ask for another one.

Join a group sponsored by the college. Students who join groups have much higher survival rates. Get involved. Attend plays, concerts, lectures, sports activities, etc. The more time you invest in activities on campus other than simply going to class, the more likely you are to survive the freshman year.

Get career planning early. Visit your campus career planning center; take a battery of vocational aptitude tests; sign up for computer assisted interactive guidance and career planning; and most importantly, see a career planning specialist to discuss your individual needs and characteristics.

Use helping services and resources offered. Use a personal counseling center. Many colleges offer academic support services, especially in the areas of math and writing skills.

Learn college study skills. Most high school students have not learned the kind of study skills they need to do well in college. Assess and improve your study skills.

Set goals and priorities at the beginning of the semester. Then develop a specific plan for achieving these goals. Baby steps are much easier than giant steps.

Develop a controlled study environment. Plan study time as part of each day's schedule. Find a place to study that is free from distractions. Try to find a place other than the bed to use for studying.

Learn effective time management. Set a realistic schedule and stick to it. Make a daily list of things to do and prioritize it. Give higher priority to items that lead to achieving semester goals. Utilize hours of peak efficiency. If you are a morning person, set a schedule that takes advantage of this. Students should schedule study breaks periodically. It is more realistic to study for 2 hours and take a 10-minute break than to attempt to study for 6 hours with no break.

Strive for a balanced life. A major time management issue for students is the balance between keeping up academically and taking care of themselves physically. In addition to scheduling study time, students should get regular exercise and an adequate amount of sleep. Fatigue affects memory and performance, so try to avoid all-nighters. A healthy diet and regular exercise helps reduce stress and improves performance.

11 Definitions of College and Financial Aid Terms

A.A. - This stands for an "associate of arts" degree, which can be earned at most two-year colleges.

A.A.S. - This refers to an "associate of applied science" degree, which can be earned at some two-year colleges.

Ability to Benefit - Applies to students who do not have a high school diploma, or its equivalent, or a GED (General Education Development Certificate). These students can still receive Federal student aid if they take a test measuring their ability to benefit from the education offered. The test must be administered independently and must be approved by the U.S. Department of Education. Students should check with their financial aid administrator for more information.

Academic Year - A period of time schools use to measure a quantity of study. For example, a school's academic year may consist of a fall and spring semester, during which a student must complete 24 semester hours. Academic years vary from school to school, and even from educational program to educational program at the same school.

Accreditation - An endorsement given to educational institutions or academic degree programs by an organization that reviews qualifications.

Accrual - An amount of money, such as interest, that continues to accumulate over the life of the loan.

ACT - American College Testing (ACT) publishes the ACT Assessment Test, commonly known as the ACT. It is a standardized, multiple-choice, and essay test used by some colleges as part of the admissions process that is administered five times a year. The ACT measures academic achievement in four areas: English, Math, Reading, and Science.

Adjusted Gross Income - Taxable income from all sources. (See your federal income tax Form 1040 EZ, line three, or Form 1040, line 31.)

Amortization - The reduction of loan balance by your monthly payments.

Annual Percentage Rate (APR) - Annual Percentage Rate represents the cost of your loan as a yearly rate. It is not the interest rate; it is calculated using your current interest rate, any fees, and for a student loan, the number of years of repayment.

Anti-Drug Abuse Act Certification - To receive a Pell Grant, you must sign a statement certifying that you will not make, distribute, dispense, possess, or use illegal drugs during the period covered by the Grant. In addition, you are also certifying that if you are convicted of a drug-related offense committed during that period, you will report the conviction in writing to the U.S. Department of Education. Your Student Aid Report (SAR) will contain this certification statement although, in some cases, your school may ask you to sign instead a separate statement it has prepared.

Assets - Items of financial worth which may include your home, business, savings and checking accounts, stock, bonds, real estate, trust funds, etc.

Audit - Attend a class without receiving credit for the class.

Average Daily Balance - The outstanding balance in a given billing cycle divided by the number of days in the cycle.

Award Letter - Issued by a college's financial aid office, this official document lists all the financial aid awarded to a student.

B.A. or B.S. - B.A. stands for "bachelor of arts," and B.S. stands for "bachelor of science." Both degrees can be earned at four-year colleges. Some colleges only grant B.A.s and others only grant B.S.s. It depends on the kinds of courses offered at the particular college.

Borrower - The person to whom a loan is made and who agrees to repay it. The borrower signs a promissory note, which serves as the formal promise to repay the loan.

Bursar - Treasurer of a college; handles tuition and fee charges, waivers, refunds.

Capitalization - If you choose to postpone paying interest on a student loan while you're in school, the interest may be capitalized, or added on to the principal amount. Capitalization increases the total amount you will repay on a student loan.

Co-Borrower (Signer) - A co-borrower is a second or additional party who receives the loan proceeds and agrees to repay the loan.

Co-Signer - A signer of a promissory note who agrees to pay the loan if the borrower defaults.

Cost of Attendance - The total amount it will cost a student to attend a particular school. This includes tuition and fees, room and board, books, supplies, transportation costs and personal expenses.

Course numbers - Numbers assigned to specific classes.

Credit-Based Loans - Loans that are made to you based on your credit worthiness as opposed to federal Stafford loans and grants which are determined through a need analysis process based mostly on the cost of education.

Credit hours - Number of hours you attend a class each week.

Debit Card - Debit cards are issued by financial institutions and allow electronic access to your bank account with a present amount of cash. When used, funds are withdrawn immediately from your account. There is no grace period and no interest to pay; transactions are rejected if the account lacks sufficient funds.

Default - Failure to pay your loan according to the terms disclosed on your promissory note. You are in default on a FFEL Program loan if your payments are more than 270 days past due or if you fail to comply with other terms of the loan.

Default Rate - The default rate is the percentage of students who took out Federal student loans to help pay their expenses but did not repay them properly.

Deferment - A period of time during repayment in which the borrower, after meeting certain criteria, is not required to make their regular monthly payments.

Degree - A certificate of completion of a course of study.

Degree plan - A specific list of required courses and electives to be completed for a degree.

Delinquent - If a payment is not received by the due date, it is considered delinquent. Delinquencies greater than 30 days are generally reported to national credit bureaus.

Direct Lending Schools - Institutions of higher education which have chosen to place all of their students' federally insured student loans through the federal Direct Lending Program. Such colleges or universities may either participate exclusively in the Direct Lending Program or allow students (borrowers) to choose the lending program from which to borrow.

Disbursement Notification - A letter that is sent to you acknowledging that your loan is approved and letting you know when the money will be sent to your school, as well as the loan amount and any fees (origination or guarantee). This makes the successful completion of the loan application process.

Disclosure Statement - A notification of the actual cost and terms of a loan, which includes the interest rate and any additional finance charges.

Dividends - Dividends are payments of part of a company's earnings to people who hold stock in the company.

Doctoral degree - The most advanced degree that can be earned.

Early Action - Permits you to apply to a college or university of your choice and receive a decision early in the senior year, well in advance of the normal spring response dates. Though you will hear early regarding your admission, you are not committed to attend and you may apply to other colleges. If you are applying for financial aid you will follow the aid application deadlines set by the institution. You are not required to make a commitment before May 1, but you are encouraged to do so as soon as a final choice is made.

Early Decision - Requires you to commit to a college or university at the time of application that, if admitted, you will enroll. You should apply under an Early Decision plan only if you know that you can make a well-reasoned, first-choice decision. Upon admission the institution will require a nonrefundable deposit well before May 1. You may apply to other colleges but may have only one Early Decision application pending at any time. Colleges will respond to requests for financial aid at or near the time admission is offered. If admitted, you must enroll unless the financial aid award is inadequate.

Education Savings Account (IRA) - This newly established (1998) IRA allows a related taxpayer to deposit up to $500 per year for each child under the age of 18. Interest earned on this type of IRA will be tax-free. In addition, withdrawals from an Education IRA will be tax-free as long as the withdrawals are used for educational expenses.

EFC (Expected Family Contribution) - An amount, determined by a formula that is specified by law, that indicates how much of a family's financial resources should be available to help pay for school. Factors such as taxable and non-taxable income, assets (such as savings and checking accounts), and benefits (for example, unemployment or Social Security) are all considered in this calculation. The EFC is used in determining eligibility for Federal need-based aid.

EFT (Electronic Funds Transfer) - The process whereby your bank sends the loan proceeds electronically to your school if the school participates in this program.

Eligible Citizen - A United States citizen, U.S. national, or resident of certain U.S. territories who qualifies to borrow under the FFEL or student lending programs.

Eligible Noncitizen - A permanent resident of the Unites States who is able to present evidence from the Immigration and Naturalization Service that he or she is in the United States for other than a temporary purpose with the intention of becoming a citizen or permanent resident.

Eligible Program - A course of study that leads to a degree or certificate and meets the U.S. Department of Education's requirements for an eligible program. To get federal financial aid, you must be enrolled in an eligible program, with two exceptions:

If a school has told you that you must take certain coursework to qualify for admission into one of its eligible programs, you can get a Direct Loan or a FFEL Program Loan (or your parents can get a PLUS Loan) for up to 12 consecutive months while you're completing that coursework. You must be enrolled at least half-time, and you must meet the usual student aid eligibility requirements.

Emergency Loan Program - These are short-term, low-interest loans that are administered by your school's financial aid office.

Exit Interview - An in-person or online counseling session with the school's financial aid office before graduation or withdrawal to review the terms and obligations of your student loan.

Federal Family Education Loan (FFEL) - A loan program authorized by the federal government in the Higher Education Act of 1965, as amended. This program includes Federal Stafford, PLUS, and Consolidation Loans. These loans are funded by lenders, guaranteed by guaranty agencies and ultimately insured by the federal government.

Federal Supplemental Educational Opportunity Grants (FSEOG) - Government grants distributed by colleges, at their discretion, to students based on need.

Fees - These are charges that cover costs not associated with the student's course load, such as costs of some athletic activities, clubs, and special events.

Finance Charge - The amount of money a lender may charge you to borrow a loan. A finance charge is also assessed by credit card companies on any unpaid monthly balances.

Financial Aid – Financial Aid in this book refers to money available from various sources to help students pay for college.

Financial Aid Package - The total amount of financial assistance a student receives, including grants, scholarships, work-study and loans, as listed in the college's financial aid award letter.

Financial Need - In the context of student financial aid, financial need is equal to the cost of education (estimated costs for college attendance and basic living expenses) minus the expected family contribution (the amount a student's family is expected to pay, which varies according to the family's financial resources).

Forbearance – A temporary postponement of principal and interest payments during which the borrower may only pay the interest on the loan. If the borrower chooses not to pay the interest, it will be capitalized at the end of the period.

William D. Ford Federal Direct Loans - Under this new program, students may obtain Federal loans directly from their college or university with funds provided by the U.S. Department of Education instead of a bank or other lender.

Financial Aid Eligibility - The difference between the total cost of attendance and the expected family contribution.

The Free Application for Federal Student Aid (FAFSA) - A standard federal form used to determine your eligibility for most types of financial aid including federal government-backed loans. The

FAFSA is typically completed early in the year and it requires income, asset, and tax information from the students and/or parents.

Freshman - A student who has completed less than 30 hours of college credit.

Full time - Twelve or more credit hours per semester for undergraduate students.

General Education Development (GED) Certificate - A certificate students receive if they've passed a specific, approved high school equivalency test. Students who don't have a high school diploma but who have a GED may still qualify for federal student aid. A school that admits students without a high school diploma must make a GED program in the vicinity of the school available to these students and must inform them about the program.

Good thru Date for Payoff - The date your payoff amount should arrive at the lender or servicing agent.

GPA - Grade point average; the average of your class grades generally based on a 4.0 scale.

Grace Period - The span of time allowed before principal repayment of a student loan must begin after graduating, leaving school, or dropping below half-time status. Some credit cards also allow a grace period before finance charges accrue on an unpaid balance.

Graduated Repayment - Offers interest-only payments for the first two or four years. It can lower initial monthly payments by as much as 40 percent, which may be especially beneficial if you're just beginning your career.

Grants - A form of financial aid, similar to scholarships that do not have to be repaid.

Gross Income - Your income before taxes and deductions.

Guarantee/Insurance Fee - A sum charged by the guaranty agency to insure a loan. The guarantee fee, sometimes called an insurance fee, is deducted from the principal amount of your loan and paid by the lender to the guaranty agency.

Guaranty Agency - An agency that insures the student loans made by lenders.

Half time - Six credit hours per semester for undergraduate students.

HOPE Scholarship Tax Credit - This tax credit is available to those taxpayers who are paying education expenses for students enrolled in their first two years of college. The Hope Tax Credit is available for up to 100 percent of the first $1,000 of qualified tuition and related expenses (i.e. tuition and fees, but not room and board and books) and 50 percent of the second $1,000. The Hope Scholarship Tax Credit is for expenses paid after December 31, 1997. It will be reduced on a pro-rata basis for those who file a single tax return with incomes of $40,000 to $50,000 and for those who file joint tax returns with incomes of $80,000 to $100,000. (Single return tax filers with incomes of $50,000 and joint return tax filers with incomes of $100,000 and above will not be eligible for a deduction.) The Hope Scholarship Tax Credit is part of the Taxpayer Relief Act of 1997.

Income-Sensitive Repayment - Establishes monthly payments based on your gross monthly income. It's particularly helpful if your income is modest. You can determine your own payment amount, within certain criteria, and can switch to another repayment plan at any time.

Internship Deferment - A period during which loan payments can be deferred (postponed) if a borrower is participating in a program required to begin professional practice or service. An internship deferment also includes participation in an internship or residency program leading to a degree or certificate awarded by an institution of postgraduate training.

Interest - The fee charged to borrow money that accrues and is paid over the life of the loan.

Interest Computation Method - Credit card companies use a different method than banks to calculate interest; the interest on an unpaid balance can be included on the next month's bill and included in the interest calculation for that payment period.

Internship - A job in a student's field of study; may be required in some academic programs and may include salary and college credit.

Investment - An investment refers to using your money to invest in something that will enable you to earn interest or dividends over time.

Junior - A student who has completed 60 to 89 college credit hours.

Late Fee - An administrative charge that a lender may assess if a student loan payment is not received within 15 days after its due date.

Lifetime Learning Tax Credit - The Lifetime Learning Tax Credit is available to people beyond the first two years of undergraduate studies, students or working U.S. citizens taking classes to improve or upgrade their job skills. It can be used for qualified tuition and related expenses (i.e. tuition and fees, but not room and board and books) paid by the taxpayers. The tax credit equals 20 percent of the first $5,000 of expenses paid by the taxpayer after June 30, 1998, through December 31, 2002, and 20 percent of the first $10,000 after January 1, 2003. The tax credits are reduced on a pro-rata basis for single return tax filers with incomes of $40,000 to $50,000 and for joint return tax filers with incomes of $80,0000 to $100,000. (Single return tax filers with incomes of $50,000 and joint return tax filers with incomes of $100,000 and above will not be eligible for a deduction.) The Lifetime Learning Credit is part of the Taxpayer Relief Act of 1997.

Liquidity - A term that refers to how quickly you can gain access to money that you invest or deposit in some kind of savings instrument.

Loan - A loan is a type of financial aid that is available to students and to the parents of students. An education loan must be repaid. In many cases, payments do not begin until the student finishes school.

Loan Consolidation - A plan that allows certain eligible lenders to pay off your existing student loans and to create one new loan. You're eligible for loan consolidation if you have loans totaling at least $5,000. You must be in repayment (or have entered your grace period) before your loans can be consolidated. You cannot be more than 90 days delinquent on any loan being consolidated. The interest rate on the consolidated loan will be 9 percent or more, depending on the interest rates of the loans consolidated. The repayment period will be from 10 to 25 years, depending on the amount to be repaid. Except for PLUS loans, most of the student loans are eligible for consolidation.

Long session - Regular fall or spring semester.

Major - A student's concentrated field of study.

Master Promissory Note (MPN) - Master Promissory Note refers to the revised promissory note that the Department of Education has authorized to be used with Stafford loans beginning July 1, 1999. It allows lenders to use a single MPN instead of requiring borrowers to sign a new promissory note for additional advances.

Master's degree - A graduate degree that usually requires two or more years of study beyond the bachelor's degree.

Merit-based Financial Aid - This kind of financial aid is given to students who meet requirements not related to financial needs. Most merit-based aid is awarded on the basis of academic performance or potential and is given in the form of scholarships or grants.

Minimum Payment - The minimum dollar amount that you must pay per month on your credit card bill.

Minor - A student's secondary field of study.

Monthly Payment - This is the amount you must pay each month on your student loan during repayment. For every $10,000 you borrow, you will pay approximately $125 per month with the standard ten-year repayment option. On a credit card, the monthly payment is usually a minimum amount due on the unpaid balance.

Multiple Disbursements - Loan proceeds that are paid in more than one check or electronic transaction. For example, a portion of a loan may go towards the first semester of school and the balance for the second semester.

Need-based Financial Aid - This kind of financial aid is given to students who are determined to be in financial need of assistance based on their income and assets and their families' income and assets, as well as some other factors.

New Borrower - A term that applies to the Stafford Loan, PLUS, or SLS programs. You're a "new borrower" under these programs if you had no outstanding (unpaid) Stafford Loans, PLUS, SLS, or consolidation loans on the date you signed your promissory note, and if your loan was either disbursed on or after July 1, 1987 or was for a period of enrollment that began on or after July 1, 1987. Once you qualify as a new borrower, the loan conditions that apply to "new borrowers" automatically apply to any future Stafford Loans, PLUS loans, or SLS loans you may receive.

Non-Direct Lending Schools - Institutions of higher education which have chosen to place their students' federally insured student loans through the Federal Family Education Loan program. In the FFEL program, borrowers choose a lender from which to borrow and the lender then obtains a loan guarantee from a state or private guarantee agency.

Non-resident - Any student who lives out of state or does not meet specific state residency requirements.

Open Admissions - This term means that a college admits most or all students who apply to the school. At some colleges it means that anyone who has a high school diploma or a GED can enroll. At other schools it means that anyone over 18 can enroll. "Open admissions," can mean slightly different things at different schools.

Origination Fee - A sum charged by the federal government on FFEL loans to offset the cost of processing the loan. The amount of the fee is deducted from the dollar amount of your loan by the lender and paid to the U.S. Department of Education.

Parent Loans for Undergraduate Students (PLUS) - These are loans under the FFEL program for parents of dependent undergraduate students. They require a credit check. The interest rate is low and repayment begins 60 days after disbursement.

Parental Leave Deferment - A period of up to 6 months when loan payments can be postponed if a borrower is pregnant, or if a borrower is taking care of a newborn or newly adopted child. The borrower must be unemployed and not attending school. To get this deferment, you must apply within 6 months after you leave school or drop below half-time status.

Payment Schedule - A summary of the terms of a loan, which includes the total principal amount, the date payment begins and the interest rate.

Payoff Balance - This is the total amount you would owe if you were to pay off your entire loan. It includes the outstanding principal plus any unpaid accrued interest.

Pell Grants - One of the largest sources of grants, Pell Grants are distributed by the federal government and are designed to help students with financial need pay for college.

Pell Grant Index (PGI) - The number that appears on your Student Aid Report (SAR), telling you about your Pell Grant eligibility. The

PGI is calculated by a standard formula that uses the financial information you reported when you applied for Federal student aid.

Perkins Loans - A campus-based, low interest loan for graduate and undergraduate students. The college acts as the lender using a limited pool of funds provided by the federal government. These loans are awarded based on exceptional financial need.

The Preliminary Scholastic Assessment Test (PSAT) - It is a two-hour test given once a year in October. As with the SAT, you receive separate math and verbal scores. In addition, the test also includes a third scored section testing English grammar. Each subject is scored on a scale of 20 to 80 and these scores are combined to create your National Merit Scholarship selection index.

Prerequisite - A course that must be taken prior to enrollment in another course.

Principal - The original amount borrowed and the amount upon which interest will be charged.

Private university - A non-state-assisted college or university that relies on private funding, tuition and fees for existence.

Promissory Note - The binding document or "promise to pay" that a borrower signs before receiving the actual loan. The promissory note includes information about the terms and conditions of the loan.

Public university - A state-assisted college or university.

Registration - Enrollment in classes.

Regular Decision - A plan in which institutions review most of their applications before notifying the majority of candidates of their admission. In this process, colleges set a deadline for completing applications and will respond to completed applications by a specified date. If you are applying for financial aid you will follow aid application deadlines set by the school. You may apply to other colleges. You will not be required to make a decision regarding enrolling before May 1.

Regular Student - One who is enrolled in an institution to obtain a degree or certificate. Generally, to receive aid from any of the programs, you must be a regular student. (For the Stafford, PLUS, and SLS programs, there are two exceptions to this requirement.

Remedial classes - Course work designed to help students bring their skills up to college level.

Repayment Period - This is the amount of time during which you repay the money borrowed plus interest.

Reserve Officers Training Corps program (ROTC) - A scholarship program wherein the military covers the cost of tuition, fees, and textbooks and also provides a monthly allowance. Scholarship recipients participate in summer training while in college and fulfill a service commitment after college.

Residency - The period during which a physician receives specialized clinical training.

Resident - A student who meets state residency requirements.

Return - Return refers to the amount of money you earn through a financial investment or savings instrument. You earn money on investments and savings instruments through interest earnings or dividends.

Risk - In reference to saving money or investing money, risk refers to the danger that the money you set aside in some kind of savings plan or investment could be worth less in the future.

Rolling admission - Term used to describe the application process in which an institution reviews applications as they are received and offers decisions to students soon after they are made. If you are applying for financial aid you will follow aid application deadlines set by the school. You may apply to other colleges and you will not be required to make a decision regarding enrolling before May 1.

Satisfactory Academic Progress - To be eligible to receive federal student aid, you must maintain satisfactory academic progress toward a degree or certificate. You must meet your school's written standard of satisfactory progress. Check with your school to find out its standard.

Savings Instrument – Savings instrument refers to any kind of savings plan or mechanism you can use to save money over time. Examples of savings instruments are savings accounts, certificates of deposits (CDs), and money market accounts.

Scholarships - Scholarships, like grants, are a form of financial aid that does not have to be repaid. These are available from many sources including community groups, schools, and private corporations. Scholarships can be awarded based on a variety of criteria including scholastic achievement, hobbies, and college majors.

The Scholastic Aptitude Test (SAT) - This is a seven-section, three-hour exam that is administered several times a year.

Selective Service Registration - If required by law, you must register, or arrange to register, with the Selective Service to receive federal student aid. The requirement to register applies to males who were born on or after January 1, 1960, at least 18 years old, are citizens or eligible noncitizens, and are not currently on active duty in the armed forces.

Semester hours - See credit hours.

Senior - A student who has completed 90 or more hours of college credit but has not received a bachelor's degree.

Sophomore - A student who has completed 30 to 59 college credit hours.

Special Conditions - Several specific circumstances that require expected 2001 income, instead of 2000 income, to be used to calculate a student's Pell Grant eligibility. These conditions accommodate certain situations where a family's financial circumstances would be worse in 2001 than in 2000. In these cases, using 2001 income would more accurately reflect the family's financial status.

Stafford Loan - Loans under the FFEL program awarded on the basis of financial need. They may be subsidized or unsubsidized. These loans can be made from a bank, credit union or other eligible lender or obtained directly from the government under the federal Direct Lending Program.

Standard Repayment - Keeps total finance charges to a minimum. It's ideal if your budget allows you to make full monthly principal and interest payments without difficulty.

Status - Indicates the condition of your student loan. Examples would include "in school" and "repayment".

Student Aid Report (SAR) - A report sent to a student by the government 4 to 6 weeks after submitting a FAFSA. The report informs the student of the Expected Family Contribution (EFC) and the financial aid for which the student is eligible. College financial aid offices use the report information to build a financial aid package for a student.

Student Loan Interest Deduction - Eligible taxpayers may deduct on their federal income tax return the amount of interest they have paid during the tax year on any qualified education loan for the period of time as defined by law. The deduction applies to interest paid after December 31, 1997, and is phased out for single taxpayers with incomes of $40,000 to $55,000 and for joint taxpayers with incomes of $60,000 to $75,000. Single return tax filers with incomes of $55,000 and joint return tax filers with incomes of $75,000 and above will not be eligible for a deduction.

Subsidized Loan - A loan on which the government pays the interest for a student while enrolled in school at least half time and during periods of grace and deferment (i.e. Subsidized federal Stafford Loan).

Summer session - A summer semester that is normally divided into two parts.

Teleregistration - Registering for your classes by phone.

Three-quarter time - Nine credit hours for undergraduate students.

Transcript - This is a list of all the courses a student has taken with the grades that the student earned in each course. A college will often require a student to submit his or her high school transcript when the student applies for admission to the college.

Tuition - This is the amount of money that colleges charge for classroom and other instruction and use of some facilities such as libraries. Tuition can range from a few hundred dollars per year to more than $20,000. A few colleges do not charge any tuition.

Unsubsidized Loan - A loan on which the borrower is always responsible for paying the interest while in school and during deferment, forbearance and grade periods (i.e. Unsubsidized federal Stafford Loan or federal PLUS Loan).

Variable Interest - Interest rates that change periodically (e.g. quarterly, annually etc.). The interest rates for federal Stafford and PLUS loans are set by the government each year and change annually on the first of July.

Wait List - Term used by institutions to describe a process in which they may initially delay offering or deny you admission, but rather extends to you the possibility of admission in the future. Colleges offer admission to wait list candidates if insufficient numbers of regularly admitted candidates accept their offers of admission.

Work-Study - Part of the federal Student Financial Assistance Program that provides part-time employment for post-secondary students who need income to help meet education costs.

W9-S - This is an IRS form which taxpayers use to certify that loans meet the definition of qualifying education debt and which allows lenders to report to the IRS the amount of interest paid on student loans as interest which qualifies for possible tax deductions.

91-day T-Bill - Refers to the auction rate determined for 91-day Treasury Bills by the public auction held by the United States Treasury Department. The interest rates for Stafford and PLUS loans are tied to the auction rates held at certain times of the year. The rate(s) can be obtained from the Treasury Department, but is also available in many newspapers including The Wall Street Journal.

12 Frequently Asked Questions

Q What can I be doing now to prepare for college?

A The first thing you should do is contact each of the schools you are interested in and request a calendar of important dates, including deadlines for admissions and financial aid applications. If you're a junior, you should start thinking about the type of school you would like to attend and start gathering information. You should also study hard. Grades will be important to the admissions office when your application is reviewed. Seniors, you will have a lot to do, but it will be worth it. Reduce your list of schools and be sure your SAT or ACT scores have been sent. Complete all of your applications, both for admissions and for financial aid. Search for scholarships and gather information on financial aid by attending financial aid workshops. And finally, get ready for school by earning money from after school and summer jobs.

Q What is financial aid?

A Financial aid is money used to pay for your education. It can be given, lent, or paid to you.

Q What types of college financial aid are available?

A There are two basic types of financial aid available to help pay for a college education: gift aid and self-help aid. Gift aid is the best, because it's money that you don't have to repay. Common types of gift aid include grants, which are usually based on a student's personal financial need, and scholarships, which are primarily awarded based on academic performance. Fellowships are another form of gift aid, based on an individual's merit, and are usually reserved for those pursuing a graduate degree. Self-help aid is money that you must earn or repay. Loans are considered a type of self-help aid, because they must usually be repaid with interest. Government run work-study programs, which provide a cash stipend in exchange for work at a specified job, are another common form of self-help aid.

Q Where can I get more information about my financial aid options?

A Federal Student Aid Information Center

1-800-433-3243 for questions about federal student aid, eligibility, form completion or correction, or publications on financial aid.

319-337-5665 Processing center for FAFSA. To find out if your FAFSA has been processed or to request a duplicate of your Student Aid Report.

1-800-730-8913 for hearing impaired - all questions.

Also, check 1-800-EDUCAID, 1-800-338-2243, for questions about Federal Family Education Loan Programs, as well as brochures and applications. Also, check www.educaid.com

Q How will I know if I am eligible?

A More people are eligible than you may think. In fact, more than 70% of students today receive some form of financial aid. After January 1 of your senior year in high school, you will submit the Free Application for Federal Student Aid (FAFSA) to start the financial aid process.

Q When do I apply to FAFSA?

A In the fall of your senior year in high school, request the FAFSA by calling 1-800-4-fed aid or apply online. You are required to submit the FAFSA if you want to receive federal financial aid, but it is also used to determine eligibility for most state and college programs. The FAFSA cannot be filed earlier than January 1, but be sure to do it as soon after that as possible.

The FAFSA is used to determine your Expected Family Contribution (EFC), which is the amount that your family will be expected to contribute to your first year of college. Once you've chosen a college, that institution's financial aid officers (FAOs) will put together a package of grants, scholarships, and/or loans that will likely make up the difference between the cost of attending and your EFC.

Q What if the package isn't enough?

A There may be some room for negotiation, particularly if your family's financial situation has changed since you submitted the FAFSA. Be sure to discuss alternative funding sources with your FAO. You can also look into lending options such as unsubsidized student loans or home equity loans that are not dependent on "need" as defined by the Department of Education.

Q How much is attending college really going to cost?

A You should check with the school that you are planning on attending. Request a catalog and review their anticipated costs.

Q When will I get aid for which I am approved?

A Typically, aid is disbursed at the start of each semester. Tuition tends to get paid directly to the school, while living expenses may get paid to you directly or wired into your bank account. Your financial aid office can give you more exact dates.

Q How often do I need to apply?

A You will need to complete the FAFSA and other forms every year that you are in school. Many scholarships and grants may be granted only for your freshman year, so be aware of this and try to plan accordingly. Reapplying will typically help families if your siblings start college while you are still in school. Last, changes in federal law may have an impact on the financial aid you receive.

Q What is the difference between "subsidized" and "unsubsidized" student loans?

A With subsidized student loans, the government pays the interest on your loan as long as you are a student (and for a grace period after you graduate).

Unsubsidized loans accrue interest from the day you get the money. (Sometimes you can defer paying the in-school interest until after you have graduated, but it will still be accrued and added to your payment amount once you leave school.)

Q What happens to my student loan if I take a semester off?

A Find out ahead of time whether you will be able to continue deferring repayment. Ask your loan servicer or your financial aid officer.

Q What is a Stafford Loan?

A You can obtain a student loan known as a Stafford loan if you attend an accredited school at least halftime. The loan is insured by a state or national guarantee agency and reinsured by the federal government. You may take Stafford loans totaling up to the limits listed below.

Undergraduates	**Dependent***	**Independent**
1st year	$2,625	$6,625
2nd year	$3,500	$7,500
3rd and 4th Year	$5,500	$10,500
Maximum Amount of Loans	$23,000	$46,000

*If you're a dependent student but your parents are denied a PLUS loan, you may borrow up to the amount that an independent student could borrow.

Subsidies: If you demonstrate financial need (according to calculations done by your college and the government, using the FAFSA and other forms), the government will pay your interest while you attend school. This is a great benefit, while you are in school; you are getting an interest-free loan. These subsidized loans can be up to the following amounts for dependent or independent students.

Undergraduate Year	**Amount Subsidized**
1st Year	Up to $2,625
2nd Year	Up to $3,500
3rd and 4th Year	Up to $5,500

Regardless of family income, you can apply for an unsubsidized Stafford loan, but you will pay interest while you're in school. Some lenders allow you to defer that interest until you're out of school, but add it to the loan principal, increasing the amount that you will owe later on.

Interest Rates. Interest rates on Stafford loans are calculated at the 91 T-bill rate plus 2.5% while you're in school (or during a grace period of deferment). Otherwise, the rate is the 91-day T-bill plus 3.1%, with a capped rate of 8.25%.

Borrower Fees. When Stafford loans are disbursed, the government deducts an origination fee of 3% and a guarantor fee of up to 1%

from the loan to offset program costs. Thus, if you borrow $1,000, your check might be for as little as $960, after you've paid your borrower fees.

Q What is a PLUS Loan?

A PLUS loans are made to parents, stepparents, or legal guardians for dependent students who are enrolled in school at least half time. Parents may borrow an amount equal to the cost of the education, less any money received from grants and any other financial aid the student receives. There are no loan limits for PLUS loans and parents do not have to demonstrate financial need, but they are subject to a mandatory credit check. Parents with a very poor credit history may not be approved for a PLUS loan. Repayment of a PLUS loan begins 60 days after you receive the money and may extend up to 10 years.

Interest Rates. Interest on the PLUS loan is one of the best credit-based loans that parents can get. PLUS loans have a variable annual rate that is based on the 52-week T-bill rate plus 3.1%, with a capped rate of 9%.

Borrower Fees. Like the Stafford loans, when PLUS loans are disbursed, the government deducts an origination fee of 3% and a guarantor fee of up to 1% from the loan to offset program costs.

Q What is the difference between grants, scholarships, and fellowships?

A Grants are awarded depending on financial need, whereas scholarships are generally given because of merit: athletic, musical, academic, or group affiliation. You can apply grants and awards to tuition but not to living expenses. Neither of these needs to be repaid, so they are the best kind of aid. Fellowships are generally awarded to graduate students, based on academic merit. They can be used for living expenses or tuition, and they do not require repayment.

Q Are all student loans alike? If so, how much?

A No. Look for lenders that provide loans that are cheaper or easier to repay, and that offer benefits for on-time payments and strong customer service. All students are eligible for some form of financial aid and are encouraged to apply regardless of their family's financial strength. Eligibility for all financial aid programs can only be determined if you apply.

Typically, applicants who apply by the school's priority deadline in March, are offered enough financial aid to meet their financial need. Financial need, or aid eligibility, is calculated by determining the Expected Family Contribution (EFC) and subtracting it from the student expense budget (cost of attendance) determined by your school.

Q What does it cost to go to college?

A Student expense budgets include the costs for tuition and fees, books, educational supplies, room and board (on-campus, off-campus, or living at home), travel/moving expenses, and personal/miscellaneous expenses. Additional costs can be approved for documented special needs. These budgets reflect average expenses paid by current students and are designed to allow a modest but adequate standard of living.

Type of Institution	Living at Home	Living On-Campus	Living Off-Campus
Public 4 Year University	$7,600	$10,600	$12,200
Community college	$5,600	n/a	$9,100
Private 4 Year University., upper rate	$19,100	$22,900	$22,100
Private 4 Year College, lower rate	$13,200	$15,400	$18,600

Q How much will my family be expected to contribute to my yearly college costs?

A Your expected Family Contribution (EFC) is calculated each year you apply for aid using a national processing formula called the Federal Methodology. The formula considers your parents' and your income and assets, your family's size, and the number of family members enrolled in college. The EFC will be the same at any college you attend.

Q What is the expected family contribution (EFC)?

A A key factor that determines whether you can receive college financial aid is your expected family contribution, or EFC. Your EFC is calculated according to a formula set by Congress. It takes into account the family's income, savings and other factors. As a general rule, parents are expected to contribute about 6% of their savings to fund your education. But, since the student presumably has fewer overall expenses, about 35% of your savings will be expected to be used to meet college costs. Your EFC will then be compared to the cost of attending the college you have selected to determine whether you are eligible for financial aid.

Q What is a school's Cost of Attendance (COA)?

A Colleges prepare a budget, technically known as their Cost of Attendance, or COA, to help students plan their educational expenses and apply for financial aid. The typical budget includes tuition, fees, books and supplies, room and board, travel, and personal and incidental expenses.

Q What is the Federal Methodology (FM)?

A The Federal Methodology, or CFM, is the formula the government uses to determine the amount of money a family is expected to contribute to their child's college education. As a result, it's a key factor in determining whether your son or daughter will qualify to receive financial aid from the government. Some colleges and universities use a different formula to determine how they dole out funds of their own. Such a formula is called an Institutional Method (IM), and can vary from one campus to the next.

Q What types of grants and loans are available?

A The following list describes the major financial aid programs and their minimum eligibility requirements:

Federal Pell Grant - Awarded to the very neediest undergraduates.

Federal Supplemental Educational Opportunity Grant - Awarded to undergraduate students with exceptional financial need.

Texas Public Education Grant - Awarded to undergraduate students showing a high financial need.

Federal and State Loan Programs - Nearly every student, whether they show financial need or not, will be able to qualify for subsidized and/or unsubsidized loan awards.

Q Are there any private sources of aid available to college students who want to study abroad?

A Many of the government-linked grant and loan programs available to students who want to study at U.S. colleges can also be used by students who wish to study abroad. In addition, there are literally thousands of privately funded scholarships and loans available to American students who would like to study in a foreign country. Requirements of these privately funded programs vary. Some require that you attend the college full-time, while others require you to take only a half-load of classes. Others dictate that you go to a particular school, or study in a particular nation. Still others require that you spend at least some of your time abroad doing public service in addition to studying.

Q What is need-based financial aid?

A Need-based financial aid is college aid that is awarded to students based on their income, as well as the income and size of their family. Need-based aid can come in a variety of forms, including loans and grants.

Q What is merit-based financial aid?

A Merit-based financial aid is aid that is awarded to you for your academic, artistic, or athletic skills. Eligibility is based on your special talents, combined with your test scores, grades and hobbies. Those criteria set merit-based aid apart from need-based aid, which is awarded according to your income level and similar factors.

Q What is the need gap for college aid?

A Colleges and the federal government have a formula for determining how much you are expected to contribute toward your teenager's college education. There's also a formula to determine how much that education will cost. The difference between those two figures is your need gap. Unfortunately, the school probably won't have the cash to fill that gap and you will be responsible to bridge it on your own. Sometimes, schools have limited funds available for financial aid, and they must determine how to best allocate the funds to their neediest students. Very few schools can afford to meet the demonstrated need of all their students, so most assume that all students and/or parents must pay a certain minimum amount, regardless of their need. Others give financial aid only to the neediest students. The college will expect the student to obtain some of the funds to fill the need gap by working, while parents may be expected to borrow money through educational loans such as the federal Parent Loan for Undergraduate Students (PLUS).

Q Is ROTC an effective way to help fund college?

A If you are interested in joining the military after going to college, it is hard to beat the financial aid program offered by the Reserve Officers' Training Corps. ROTC provides students with money for college while they are in school. In exchange, the student must make a commitment to join the armed services, after graduating. Upon graduation, students become commissioned officers in the Navy, Army, Air Force or Marine Corps. ROTC scholarships pay for tuition, fees, and books for all 4 years of college. Smaller programs also are available for students who want to make lesser commitments.

Q What type of financial aid is available to students who want to attend medical school?

A There are several government and private sources of college aid for students who want to study medicine. Some students can "piggy-back" this aid with money from programs geared toward specific groups, such as women and minorities, to help meet the high cost of going to medical school. The financial aid office at the school you plan to attend can tell you about the different types of help that are available.

Q What schools am I eligible to attend with a student loan?

A You can use your student loan to attend most universities, colleges (including junior colleges) or vocational schools (including business or trade schools or technical institutions) that limit enrollment to high school graduates (or persons with equivalency certificates or who are beyond the age of compulsory school attendance) and that are accredited by a nationally recognized accrediting agency or association approved by the federal government.

Q What expenses do student loans cover?

A Most expenses related to the cost of an education are allowable: tuition, fees, books, supplies, room and board, and even transportation to and from school.

Q If I receive a scholarship, grant, or other financial assistance, am I still eligible for a student loan?

A Yes, however, the total amount of the student loan might be affected based upon other aid you are receiving. After you complete the "Needs Analysis Form FAFSA" (which you can obtain from the school you plan to attend) and return it as instructed, you will have a better idea of the specific amounts available.

Q Do I have to re-apply for a student loan each year?

A Yes, you will need to complete a "Needs Analysis Form," FAFSA, and loan application each year you borrow money. However, you are not required, in any way, to continue borrowing each year. For Stafford Loan borrowers as long as you remain enrolled in school at least half time, you do not need to begin repaying your loan.

Q When must the loan be repaid?

A Repayment of the loan begins about six months after graduation, or six months after the student becomes enrolled for less than half time. PLUS Loan repayments begin within 60 days of the final disbursement. Certain repayment deferments are available if you qualify. While most borrowers repay their loans as agreed, there are severe penalties for non-payment. Defaulting on a loan will be reported to the credit bureau and will prevent you from receiving additional financial assistance and will probably prevent you from receiving other types of loans for cars, credit cards and even a mortgage loan. If you default on the loan, the Internal Revenue Service may hold your tax refund and your employer can garnish your wages.

Q How long do I have to repay the loan?

A Loan repayment terms can be up to ten years, but a minimum monthly payment of $50 is required. Loan consolidation and longer terms are available under certain circumstances.

Q How do I apply for a student loan? Who will my lender be?

A Applying for a loan is a simple process:

1. You must complete a "Needs Analysis Form" (called the Free Application For Federal Student Aid or FAFSA Form) that you can obtain from the school you plan to attend or the Department of Education.

2. Next, you should contact the financial aid office of the school you plan to attend and request that they send you the school's financial aid package.

3. After completing the "Needs Analysis Forms FAFSA" and any additional requirements of the school you plan to attend, mail these forms back according to the specific instructions. The school's financial aid officer will notify you of the aid and loans you may qualify for via an awards letter.

4. When the loan has been awarded, the school will notify you. In some cases the school will send you a Stafford and/or PLUS Student Loan Application partially completed for you. If you do receive these forms you must use them. If you are not mailed these forms, then contact an eligible student loan lender to obtain the proper application form.

5. On your application form that you return to your school you will be asked to indicate who your lender will be. Federal law specifies several categories of lenders that are eligible to originate loans under the Federal Family Education Program. They include any National or State chartered bank, savings and loan association, credit union, pension fund, insurance company, or the Department of Education. In the state of Texas, the Texas Higher Education Coordinating Board is eligible to make loans for enrollment at most Texas colleges and universities.

Q Can I earn other income from working while going to school and receiving the loan?

A Yes, but again, the amount of the student loan might be affected by the amount of income the student is earning while in school. The "Needs Analysis Form," FAFSA, will help answer this question for you, including any work/study programs for which you may qualify.

Q Will my student loan help me establish credit?

A Yes, as with most loans, your student loan will be reported to a national credit bureau and an excellent payment record will help establish a good credit rating.

Q After I am approved for the loan, when do I receive the money?

A The school you are attending will specify the dates that the loan proceeds should be disbursed. Most loans are funded in two installments at the beginning of each semester and are usually disbursed directly to the school according to their procedures. Your school can give you more information about their policies.

Q Are all lenders that participate in the FFELP alike?

A No, although the interest rate and terms are essentially the same for all lenders, the quality of service and opportunities for savings differ. Some banks offer repayment options and savings programs developed by Sallie Mae, the nation's leading provider of education loan services.

Q What kind of financial aid does the Department of Education have to offer?

A Federal student aid includes two types, gift aid and self-help aid. You don't have to pay back gift aid, and the self-help aid programs allow you to earn money or to borrow money for school. You'll have to repay any money you borrow.

The gift aid programs are the Federal Pell Grant Program and the Federal Supplemental Educational Opportunity Grant (FSEOG) Program. These grants are available only to students who do not yet have bachelor's degrees.

Federal Stafford Loans are the Department's major form of self-help aid for students. Federal PLUS Loans are for parents of students.

Q What criteria do I have to meet to get federal student aid?

A Basically, to receive aid from the Department of Education, you must:

- Have financial need,
- Have a high school diploma or a General Education Development (GED) certificate or pass a test approved by the Department of Education,
- Be enrolled as a regular student (that is, you have to be working toward a degree or certificate) in an eligible program,
- Be a U.S. citizen or eligible noncitizen,
- Have a valid Social Security Number,
- Register with the Selective Service if required, and
- Maintain satisfactory academic progress once in school.

Q How do I apply for student aid?

A Applying for federal student aid is easy; you don't have to fill out form after form. All you have to do is complete a Free Application for Federal Student Aid (FAFSA) and mail it to the address indicated on it. Or you may be able to apply electronically.

You can get a FAFSA (or a Spanish version of the FAFSA) from your high school, the college or career school you plan to attend, or the Federal Student Aid Information Center. The address and phone number are:

Federal Student Aid Information Center
P. O. Box 84
Washington, D.C. 20044
1-800-4-FED-AID

Q What kind of information do I need to put on my application?

A The FAFSA asks for your family's financial information. When you fill out the FAFSA, you'll need your parents' latest income tax return. If you filed a return, you'll need yours too. Referring to the tax forms makes it easier to answer the FAFSA questions. Bank statements, W-2 forms, and business/farm records might also be helpful. Save all of the forms you refer to when completing the FAFSA because you might need them later. You will also need your drivers' license number. It's a good idea to keep a photocopy of your completed FAFSA in case there is a discrepancy. If the information is incorrect, you won't get any aid until you correct it.

On the FAFSA, you can list schools you're interested in attending, and those schools will get the results of your FAFSA after it's processed. You aren't required to list any schools on your FAFSA, but if you do so, the school you decide on can deliver your aid faster. A school that participates in at least one of the federal aid programs will have a Title IV Institution Code. For each school that interests you, list its code in the appropriate section of the FAFSA. You can get Title IV codes from a college or career school financial aid office, your high school, or your local public library.

Q What is "need"?

A College financial assistance is connected to need; and need, the way colleges define it, has two sections.

1. What college costs - One kind of cost is the cost of going to school: tuition, books, supplies, and fees. The other one is the cost of living while you go to school: food, housing, transportation, and personal expenses.
2. What you and your family can afford to contribute - Colleges consider your family's income and expenses. For example, if your family has a low income, colleges and universities will figure you need a lot of financial assistance. And what if you have a lot expenses? For example, your dad makes a good salary, but there are four kids, one already in college, one in daycare, and grandmother needs help with long-term care. All these concerns will be taken into consideration. Colleges will conclude that you require substantial financial assistance.

Q What is "aid"?

A After a college or university decides what you need, it decides how much aid to offer, and what kind. There are four main types:

1. Grants: Given based on need; they do not have to be repaid.
2. Scholarships: Given based on need or another basis (a good PSAT, SAT, ACT score); they do not have to be repaid.
3. Loans: From the government, banks, or colleges; these must be repaid after you graduate or leave school.
4. Work-study jobs: Distributed by the colleges; these let you earn money while working on campus.

Q What is the Hope Scholarship credit?

A The Hope credit is only for the first two years of college. It is equal to 100% of the first $1,000 in tuition and fees, and 50% of the next $1,000, for a maximum of $1,500. This credit is phased out between incomes of $40,000 to $52,000 for singles and $82,000 to $102,000 for joint filers. It is a per-child credit. When originally authorized in 1997, the government expected some 5.9 million students annually to use the credit, resulting in $5.6 billion in savings. The government estimated that 17% of all undergraduate students or their families used the Hope credit in 1999-2000. The credit was equal to 20% of tuition and fees charged to students claimed as dependents by their families and 30% of those costs charged to students paying their own way.

Q What is College tuition deduction?

A Parents cannot use this deduction with the Hope or Lifetime Learning credit for the same student in the same year. But, if the parents' income is too high for those credits, parents may get some relief here. For 2002 and 2003, if the parents' income is less than $65,000 ($130,000 on a joint return), they may deduct as much as $3,000 per year for qualified education expenses paid, above the line. That is, they can claim the deduction even if they do not itemize. For a family in the 27% bracket, being able to claim the deduction means your tax bill will drop by $810. In the 30% bracket, the deduction is worth $900 in reduced federal taxes.

Q What about my Student loan interest?

A The government gives you deductions and credits while you are in college. They will give you tax breaks to save for college. Also, the government now gives you breaks after you graduate. Your student loan interest can now be deducted regardless of the age of the loan or when it was paid. The old 60-month limit for deductibility is history.

Disclaimer - Students in search of financial aid to pay for college may find some of the links and data useful. This information is provided for informational purposes only. Inclusion does not constitute endorsement. Nobody can guarantee the accuracy or the timeliness of data found at any of the links or any data within the book. It is critical that students verify information, eligibility, and dates, particularly where deadlines are concerned.

About the Author

Brenda Smith received a B.B.A. in Business Administration from Dallas Baptist University as well as an Associate's degree in Liberal Arts from Tarrant County College.

For six years, she investigated libraries, foundations, and associations along with interviewing professors, teachers, counselors, and students to discover the best method to win scholarships. During the past few years, she has coordinated many searches with extraordinary outcomes for students that needed scholarships and low or no interest loans for college. Those searches finished with incredibly high rates of success and unbelievable financial results for each candidate.

According to Brenda, her reason for writing *The Best Kept Secrets for Winning Scholarships* was that she wanted to share her secrets with students who needed financial assistance because "everyone deserves an education." Her seminars, one-on-one counseling, and articles on how to find free money for college have helped numerous students achieve their dreams of going to college. Apart from helping students, her interests include family, church, writing, travel, and classic Volkswagens. Brenda currently resides in Arlington, Texas with her husband, their Chihuahua, and a 1973 VW Bug.

www.bestkeptsecretsforwinning.com